Construction Project Monitoring and Evaluation

This book will provide readers with an in-depth theoretical awareness and practical guidance on the implementation of an effective monitoring and evaluation (M&E) system to ensure construction projects meet approved quality, cost, time and social sustainability objectives.

The authors discuss the drivers, challenges, determinants and benefits of effective M&E implementation together with the theories and models underpinning construction project M&E practices. Further, a comparative overview of M&E practices in developed and developing countries is presented to elucidate the best practices. The book first conceptualizes M&E as a five-factor model comprising stakeholder involvement, budgetary allocation and logistics, technical capacity and training, leadership, and communication. It then presents an M&E case study on the Ghanaian construction industry before expanding on the idea of M&E systems as an effective tool for project performance and in optimizing a project's contribution to society and the environment.

The book further provides guidance on M&E practice for construction project managers, investors, professionals, researchers and other stakeholders and is therefore of interest to those in architecture, construction engineering, planning, project management and development studies.

Dr. Callistus Tengan is a Senior Lecturer in the Department of Building Technology at Bolgatanga Technical University, Ghana and a Professional member of the Ghana Institution of Surveyors (GhIS). He is a Postdoctoral Research Fellow in the Department of Construction Management and Quantity Surveying of the University of Johannesburg, South Africa.

Professor Clinton Aigbavboa is the Director of the CIDB Centre of Excellence & Sustainable Human Settlement and Construction Research Centre, Faculty of Engineering and the Built Environment, University of Johannesburg, South Africa.

Professor Wellington Didibhuku Thwala is the Director of the South Africa Research Chair in Sustainable Construction Management and Leadership in the Built Environment, Faculty of Engineering and the Built Environment, University of Johannesburg, South Africa.

Construction Project Monitoring and Evaluation
An Integrated Approach

Callistus Tengan, Clinton Aigbavboa, and Wellington Didibhuku Thwala

Routledge
Taylor & Francis Group

LONDON AND NEW YORK

First published 2021
by Routledge
2 Park Square, Milton Park, Abingdon, Oxon OX14 4RN

and by Routledge
52 Vanderbilt Avenue, New York, NY 10017

Routledge is an imprint of the Taylor & Francis Group, an informa business

© 2021 Callistus Tengan, Clinton Aigbavboa and Wellington Didibhuku Thwala

British Library Cataloguing-in-Publication Data
A catalogue record for this book is available from the British Library

Library of Congress Cataloging-in-Publication Data
Names: Tengan, Callistus, author. | Aigbavboa, Clinton, author. | Thwala, Wellington, author.
Title: Construction project monitoring and evaluation : an integrated approach / Callistus Tengan, Clinton Aigbavboa, Wellington Thwala.
Subjects: LCSH: Building--Superintendence. | Project management.
Classification: LCC TH438 .T46 2021 (print) | LCC TH438 (ebook) | DDC 624.068/4–dc23
LC record available at https://lccn.loc.gov/2020050908
LC ebook record available at https://lccn.loc.gov/2020050909

ISBN: 978-0-367-68529-4 (hbk)
ISBN: 978-0-367-68532-4 (pbk)
ISBN: 978-1-003-13797-9 (ebk)

Typeset in Goudy Oldstyle Std
by KnowledgeWorks Global Ltd.

Contents

PART III
Communication and leadership in monitoring and evaluation   85

Figures

Tables

Maps

Preface

Undertaking an effective monitoring and evaluation to achieve success and ensure sustainability of construction projects has been a universal worry to project implementers such as donors, funders, development partners and professionals in the construction industry. While it is considered inevitable to ensure that projects meet the approved quality standards, cost, time while observing the best construction practices and meeting the social sustainability objectives, the influence of a robust and an effective monitoring and evaluation (M&E) system in project implementation would best complement team efforts in achieving success and sustainability in project delivery. The current book will therefore provide readers with an in-depth theoretical awareness on the implementation of M&E. Also discussed in the book are the drivers, implementation challenges, determinants and benefits of effective M&E implementation together with the theories and models underpinning construction project M&E practices. Further, a comparative overview of M&E practice in both developed and developing country context is presented to elucidate the best practices. The book investigates and conceptualizes M&E as a five-factor model comprising stakeholder involvement, budgetary allocation and logistics, technical capacity and training, monitoring and evaluation leadership, and monitoring and evaluation communication from an extant literature review. Likewise, an M&E Delphi case study on the Ghanaian construction industry is presented. The book expands the idea of M&E as an effective tool for project performance and to optimize the implemented projects to societal good and contribute positively to the environment.

The book further provides guidance on M&E practice for construction project donors, funders, professionals, researchers and other stakeholders alike. The book is therefore of interest to researchers and policy-makers in architecture, planning and management of projects, donors and development studies.

Callistus Tengan Ph.D.
Clinton Aigbavboa, Ph.D.
Wellington Didibhuku Thwala, Ph.D.

Part I

Performance management and measurement

1 Project management

1.1 Abstract

Project management is a broad function that entails project performance measurement and assessment. Therefore, project activities, resources and processes in the construction industry require to be assessed and measured effectively to ensure that success factors are achieved. However, little attention has been given to monitoring and evaluation (M&E) as a performance management and measurement tool. The first chapter of this book is dedicated to contextualize M&E in the project management literature as a performance management and measurement tool. Technologies such as drones and unmanned aerial vehicles (UAVs), building information modelling (BIM) and geographical information systems (GIS) have been discussed and the advantage of adapting them by the construction industry to effectively and efficiently monitor and evaluate the construction process and activities presented.

1.2 Introduction

The construction industry in many economies serves a critical role in its development trajectory. It is an industry that brings together several other industries, i.e. automobile, manufacturing, etc., under its umbrella, hence very pivotal in the development of other sub-sectors and industries in the economy. The employment generation and creation of the industry are great since it is more labor-intensive as compared to other industry (Kim, Kim, Shin, & Kim, 2015, p. 1534). Its ultimate role in the contribution to the gross domestic product (GDP) has gained recognition in most budget readings of most countries across the globe. According to Mazhar & Arain (2015, p. 434), the construction industry's contribution to the global share of the GDP is estimated around 15%. Considering the potential contribution of the industry to economic development, it is imperative to ensure that the inputs (resources), activities and processes are managed efficiently and effectively. This can be achieved via the effective implementation of monitoring and evaluation (M&E) which seeks to manage and measure performance of the project throughout the lifecycle of the project (Tengan & Aigbavboa, 2016; Tengan & Aigbavboa, 2018). This chapter therefore, seeks to elevate M&E in the project management discourse while presenting innovative and emerging technologies in assessing and measuring project performance.

1.3 Project performance management

In construction, project management has been defined by Fewings (2005, p. 11) as the "...planning, monitoring and control of all aspects of a project and the motivation of all those involved to achieve the project objectives on time and to cost, quality and performance". Further, the Project Management Institute (PMI, 2013) describes project management as the art of directing and coordinating human and material resources through the life of a project using contemporary management techniques in order to achieve set goals such as scope, cost, time, quality and participant satisfaction. The impact of PM practice on the outcome of projects has made the discipline seen as a key competence and the most sought-after career choice for many organizations in the world today both in the private and public sectors. PM skills and competencies are therefore necessary for project actors in the construction process to effectively deal with all conceived challenges and project risk ensuring that projects are initiated and closed rightly. PM has also been defined by many scholars and researchers in literature providing characteristics as they may apply to the specific fields and industry. PM practices employ performance measurement and assessment techniques such as M&E, to ensure project cost, quality and time are achieved.

In the literature, project performance management encompasses performance assessment and measurement which most often have been used to mean the same. It should, however, be noted that there exist some apparent differences between the two terminologies. According to Amaratunga et al. (2002) and Hwang & Lim (2013) performance measurement provides a set of indicators for measuring the outcome/output of daily operations, whereas performance assessment is the process of evaluating the actual performance against the set standard (Myeda & Pitt, 2012). In assessing and measuring project performance, a number of tools or models are employed in the construction industry to ensure effective M&E. This includes the Total Quality Management (TQM), the Balance Scorecard (BSC), the Key Performance Indicators (KPI), the Just-in-Time (JIT) model and the Six-Sigma. The purpose of each model, tool or system is to improve on the current performance in project activities. Similarly, the M&E tool or system ensures that set targets (KPI) are achieved, hence improving performance. M&E integrates various aspects of the project life cycle such as inputs, activities, output and outcomes to achieve success.

1.4 Monitoring and evaluation as a performance management tool

A successful project is expected to meet three key performance indicators. The three key indicators, also referred to as the triple constraints of projects, are described as project cost, quality and schedule (Rahschulte & Milhauser, 2010).

Several performance indicators have evolved over the years. Pinto and Slevin (1988) identified project schedule, budget, performance and client satisfaction as the parameters with which to measure the success of a project. Several other studies have mentioned project success factors to include environmental factors, health and safety, effective project management and governance practices (Mirza, Pourzolfaghar & Shahnazari, 2013; Akanni, Oke & Akpomiemie, 2015; Das & Ngacho, 2017). However, to ensure success of project, thus meeting the agreed performance measures for the project, monitoring, evaluation and controlling of the project have been topical in the field of project management (Ile et al., 2012; Zhao, Mbachu & Domingo, 2017). Igbokwe-Ibeto (2012) adds that due to the tedious nature of construction projects, it is imperative to monitor and evaluate projects in order to achieve success. Unfortunately, M&E as a performance measurement and assessment tool is missing in the literature. The role played by M&E in contributing to the broader spectrum of successful construction project delivery is essential and, as such, should be recognized across sectors of developing economies (Tengan & Aigbavboa, 2018), particularly in the construction industry. In the current economic constraint, particularly in developing nations, project M&E has become topical in demonstrating accountability and project impact (Barasa, 2014). M&E prompts the conditions under which projects are likely to succeed or falter and can serve as an early warning tool for potential problems. It can also lead to ideas for potential remedial action. The combined effect of M&E on a Construction Project (CP) will result in rich knowledge generation, construction programme improvement, accountability, transparency, resource allocation, advocacy and impact assessment. The above is brought to bear through the diversities of stakeholders that are usually involved in the monitoring and evaluation of every CP. The lack of integration of M&E models and practice in the Construction Industry (CI), particularly in Ghana, into mainstream CP management has accounted for many CP failures in project delivered according to time, cost, quality and meeting set health and safety (H&S) standards. This book attempts to fill this gap.

1.5 Emerging trends in project monitoring and evaluation

Monitoring and evaluation generally has been largely recognized as a human-centered activity. In the construction industry, project teams (consultants) consisting of professionals such as engineers, architects and quantity surveyors are contracted and tasked with the responsibility of ensuring that designs and other development objectives are implemented within the framework to achieve the objectives of the project. Personal site visits are undertaken by stakeholders to ascertain progress and sometimes the quality of projects. Projects also require the submission of photographs as an evidence of the progress made. Other traditional approaches have been employed off site to verify and validate other key performance indicators such as cost and laboratory testing to validate quality.

With the emergence of technology, stakeholders on a project are able to work remotely and still ensure the right things are done and are also able to access any relevant information required for the purpose of evaluation and decision making. Emerging/emerged technologies that are facilitating M&E in complex project environments are the use of drones or unmanned aerial vehicles (UAVs), Geographic Information Systems (GIS) and Building Information Modelling (BIM). As confirmed in the literature, the construction industry has been slow in embracing new technologies as compared to others such as manufacturing, even though the long-term advantages are well known. PwC asserts that the use of drones or UAVs in a construction project offers an unparalleled record of all activities, cuts planning and survey costs, increases efficiency and accuracy and eliminates disputes over the status of a project at any given point in time in the life cycle of the project.

1.5.1 Drone and unmanned aerial vehicles (UAVs)

A drone, also referred to as an unmanned aerial vehicle (UAV), is a flying robot controlled from a remote unit and due to its intelligent software programming, it can manage all things in air. The high inefficiencies, poor safety, project delays and cost overruns reported in the project management literature can potentially be improved with the deployment of drones in the monitoring and evaluation process. Drones have the potential to increase impact of the M&E process through data acquisition, processing and management for projects. Managing a construction project is no small task. From tracking site progress and monitoring safety and security to overseeing subcontractors and keeping stakeholders informed, there is almost no end to the amount of coordination you face on any given day, hence the need for project managers and M&E teams on projects to include drones to their toolkit. On a construction site, drones can assist with pre-construction site review, aerial surveying and mapping, measurement of excavation depths and material stockpiles, monitoring and documenting job site progress, productivity and inspecting work that is difficult or dangerous for human inspectors to reach. Not only can drones ensure efficient utilization of project resources, but they can also give your team a rich set of data for more informed communication and decision making, i.e. data taken by drones can be used to assist a design team in understanding the project site orienting structures and locating utilities.

1.5.2 Building information modelling (BIM)

Building Information Modelling (BIM) is an innovative technology, a repository of digital information, a modeling technology and a global digital technology which enhances the management of project information and the construction process. It helps in the creation and maintenance of an integrated collaborative database of multi-dimensional data concerning the design, construction and operations of projects, with the aim of improving collaboration between stakeholders and reduce the time needed for documentation of the project and producing more predictable project outcomes (Abanda et al., 2015; Fazli et al., 2014; Olawumi & Chan, 2019; Sampaio, 2015). The impact of BIM on the construction process is enormous, ensuring that BIM provides "single, non-redundant, inter-operable information repository" capable of supporting every stage, process and functional units in a construction project (Olatunji, Olawumi & Ogunsemi, 2016; Olawumi & Ayegun, 2016; Olawumi, Akinrata & Arijeloye, 2016). An integrated M&E approach will be enhanced with the introduction and adoption of BIM in the construction process. The call for relevant competencies in BIM application on projects needs immediate attention.

1.5.3 Geographic information system (GIS)

A GIS is a set of tools comprising hardware, software, data and users. This set of tools allows for capturing, storing, managing and analyzing digital information or data and also making graphs and maps and representing alphanumeric data (López Trigal, 2015). In the simplest terms, GIS merges cartography, statistical analysis and database technology. In M&E, data or information on project progress, quality and cost are collected, analyzed and decisions are taken based on these data. In recent times, project delivery objectives are more aligned to ensuring sustainability indicators such as health and environment. This requires the adoption of an approach or system and technologies that will facilitate the collection of such information for decisions to be taken. Towards achieving that, a GIS readily comes to mind. GISs' application in the construction industry is limited, compared to its application in addressing land and natural resource management problems and environmental issues. In the construction industry, GIS has been applied in construction safety planning to understand the execution sequence in safety planning (Bansal, 2011). While the merging of M&E and GIS into a single assessment tool to display useful information to support successful project outcomes is acknowledged as a challenge, the complement of the two activities can be seen in their distinct applications. M&E focuses on measuring the changes and the outcomes occurring over the project duration, while GIS is concerned with identifying where these outcomes occurred.

1.6 Professional ethics in the construction industry

The construction industry's activities and processes influence and affect human lives. It is also an industry that brings professionals such as clients, engineers, architects and quantity surveyors, as well as industry stakeholders together to relate. This relationship must be professional and guided by ethics (Martin & Schinzinger, 1996). Professionalism has been defined to mean the exercise of a body of unique and expert knowledge (Fellows, 2003). The need, therefore, for professionals in the construction industry to acquire relevant knowledge set is imperative for professional practice. The knowledge set for professionals goes beyond academic laurels to continuous professional development (CPD) to inculcate contemporary issues in practice. Similarly, Rosenthal and Rosnow (1991: 231) inform, "...ethics refers to the system of moral values by which the rights and wrongs of behavior... are judged". Ethics are usually personal and reciprocal in application; thus, you treat others the way you want others to treat you.

Ethics and professionalism are critical for the sustenance of the construction industry. Adherence of ethics and professional practice has a direct impact on the success of the industry performance. Abdul-Rahman et al. (2010) posits that professional ethics are a pre-requisite to reaching sustained and acceptable quality in construction. However, Johnson (1991) argued that professionals tend to place premium on their obligation towards their clients more than their responsibility towards others, while Coleman, as early as 1998, had caused to lament on the non-adherence to ethical standards in the late 1960s. The practice of M&E must be undertaken within a framework of ethics and professionalism to ensure information generated is reliable and sufficient for decision making. Professionals operating within the construction business environment such as the quantity surveyors, engineers and architects, are guided by approved codes of ethics and professional practice. Nonetheless, Vee and Skitmore (2003) aver the need to complement the efforts of codes of ethics with ethics officer who will ensure sanctions for breach are enforced.

1.7 Aim of the book

Undertaking an effective monitoring and evaluation to ensure success and sustainability of construction projects has been a universal worry to project implementers such as donors, funders, development partners and professionals, in the construction industry. While it is considered inevitable to ensure that projects meet the approved quality standards, cost, time and, at the same time, observing the best construction practices and meeting the social, economic and environmental sustainability objectives, the influence of a robust, effective and integrated M&E system during project implementation would best complement team efforts in achieving the desired success and sustainability during project delivery. The current book aims to provide readers with an in-depth theoretical awareness on the integrated approach to the implementation of M&E. Also discussed in the book are the drivers, implementation challenges, determinants and benefits of effective M&E together with the theoretical and conceptual underpinnings of construction project M&E practice. Further, a comparative overview of M&E practice in the context of the developed and developing countries is presented to elucidate the best practices. The book investigates and conceptualizes M&E as a five-factors-integrated model comprising stakeholder involvement, budgetary allocation and logistics, technical capacity and training, monitoring and evaluation leadership and monitoring and evaluation communication from an extant literature review and the Delphi study findings. The book expands the idea of M&E as an effective performance measurement and assessment tool for project management practice and to ensure optimization of the implemented projects to societal good and contributes positively to the environment. The book serves as a resource guide for construction professionals, researchers and other stakeholders alike. The book should interest researchers and policy-makers in architecture, planning and management of projects, donors and development studies.

Summary

The first chapter of this book briefly contextualizes M&E in the project management literature as a performance management and measurement tool. Technologies such as drones and unmanned aerial vehicles (UAVs), building information modelling (BIM) and geographical information systems (GIS), were discussed as innovative and emerging technologies, upon their adoption and integration in the construction industry, will enhance the effective M&E of the construction process. The need for stakeholders to be guided by professionalism and ethics during the M&E of the construction process and activities in the industry is critical. The next chapter provides an overview of monitoring and evaluation research.

References

Abanda, F. H., Vidalakis, C., Oti, A. H. & Tah, J. H. M. (2015). A critical analysis of Building Information Modelling systems used in construction projects. *Advances in Engineering Software*, 90, 183–201. https://doi.org/10.1016/j.advengsoft.2015.08.009

Abdul-Rahman, H., Wang, C. & Yap, X. W. (2010). How professional ethics impact construction quality: Perception and evidence in a fast developing economy. *Scientific Research and Essays*, 5(23), 3742–3749.

Akanni, P. O., Oke, A. E. & Akpomiemie, O. A. (2015). Impact of environmental factors on building project performance in Delta State, Nigeria. *HBRC Journal*, (11)1. 91–97, ISSN 1687-4048, https://doi.org/10.1016/j.hbrcj.2014.02.010.

Amaratunga, D., Baldry, D., Sarshar, M. & Newton, R. (2002). Quantitative and qualitative research in the built environment: Application of 'mixed' research approach. *Emerald Group Publishing Limited*, 51(1), 17–31.

Bansal, V. K. (2011). Application of geographic information systems in construction safety planning. *International Journal of Project Management*, 29(1), 66–77. https://doi.org/10.1016/j.ijproman.2010.01.007

Barasa, R. M. (2014). *Influence of monitoring and evaluation tools on project completion in Kenya: A case of Constituency Development Fund projects in Kakamega County, Kenya.* Kenya: University of Nairobi.

Coleman, J. W. (1998). *The criminal elite, understanding white-collar crime.* New York: St. Martin's Press.

Das, D. & Ngacho, C. (2017). Critical success factors influencing the performance of development projects: An empirical study of Constituency Development Fund projects in Kenya. *IIMB Management Review*, (29)4, 276–293, ISSN 0970-3896, https://doi.org/10.1016/j.iimb.2017.11.005.

Fazli, A., Fathi, S., Enferadi, M. H., Fazli, M. & Fathi, B. (2014). *Appraising effectiveness of Building Information Management (BIM) in project management.* CENTERIS 2014 – Conference on ENTERprise Information Systems/ProjMAN 2014 – International Conference on Project MANagement/HCIST 2014 – International Conference on Health and Social Care Information Systems and Technologies. Procedia Technology 16, 1116–1125.

Fellows, R. (2003). *Professionalism in construction: Culture and ethics.* CIB TG 23 International Conference, October 2003, Hong Kong.

Fewings, P. (2005). *Construction project management: An integrated approach.* USA and Canada: Taylor & Francis.

Hwang, B. G. & Lim, E. S. J. (2013). Critical success factors for key project players and objectives: Case study of Singapore. *Journal of Construction Engineering and Management,* 139(2), 204–215.

Igbokwe-Ibeto, C. J. (2012). Issues and challenges in local government project monitoring and evaluation in Nigeria: The way forward. *European Scientific Journal,* 8(18).

Ile, I. U., Eresia-Eke, C. & Allen-Ile, C. (2012). *Monitoring and evaluation of policies, programmes and projects.* Hatfield, Pretoria: Van Schaik.

Johnson, D. G. (1991). *Ethical issues in engineering.* New Jersey, USA: Prentice Hall.

Kim, S., Kim, J. D., Shin, Y. & Kim, G. H. (2015). Cultural differences in motivation factors influencing the management of foreign laborers in the Korean construction industry. *International Journal of Project Management,* 33(7), 1534–1547. doi:10.1016/j.ijproman.2015.05.002

López Trigal, L. (2015). *Dictionary of applied and professional Geography. Territory analysis, planning and management terminology.* Leon, Spain: University of Leon.

Martin, M. W. & Schinzinger, R. (1996). *Ethics in engineering* (3rd edn.). New York: McGraw-Hill.

Mazhar, N. & Arain, F (2015). Leveraging on work integrated learning to enhance sustainable design practices in the construction industry. International Conference on Sustainable Design, Engineering and Construction, *Procedia Engineering,* 118, 434–441.

Mirza, M. N., Pourzolfaghar, Z. & Shahnazari, M. (2013). Significance of scope in project success. *Procedia Technology,* 9, 722–729. doi:10.1016/j.protcy.2013.12.080

Myeda, N.E. and Pitt, M. (2012). Understanding the performance measurement system (PMS) for facilities management (FM) industry in Malaysia. PMA 2012 Conference, Cambridge.

Olawumi, T. O. & Chan D. W. M. (2019). Building information modelling and project information management framework for construction projects. *Journal of Civil Engineering and Management,* ISSN 1392-3730/eISSN 1822-3605, 25(1), 53–75. https://doi.org/10.3846/jcem.2019.7841

Olatunji, S. O., Olawumi, T. O., & Ogunsemi, D. R. (2016). Demystifying Issues Regarding Public Private Partnerships (PPP). *Journal of Economics and Sustainable Development - The International Institute for Science, Technology, and Education (IISTE),* 7(11), 1–22

Olawumi, T. O., & Ayegun, O. A. (2016). Are quantity surveyors competent to value for civil engineering works? Evaluating QSs' competencies and militating factors. *Journal of Education and Practice,* 7(16), 1–16.

Olawumi, T.O., Akinrata, E.B., & Arijeloye, B.T. (2016). Value Management- Creating Functional Value for Construction Projects: An Exploratory Study. *World Scientific News, WSN,* 54, 40–59.

Pinto, J. K. & Slevin, D. P. (1988). Project success: Definitions and measurement techniques. *Project Management Journal,* 19(1), 67–73.

Project Management Institute (PMI) (2013). *Managing change in organizations: A practice guide.* UK: PMI.

Rahschulte, T. J. & Milhauser, K. (2010). Beyond triple constraints – Nine elements defining project success. In: *Project Management Institute.* North America, Washington, DC. Newtown Square, PA: Project Management Institute.

Rosenthal, R., Rosnow, R. L. (1991). *Essentials of behavioral research: Methods and data analysis* (2nd edn.). Boston, MA: McGraw-Hill.

Sampaio, A. Z. (2015). The introduction of the BIM concept in civil engineering curriculum. *International Journal of Engineering Education*, 31(1B), 302–315.

Tengan, C. & Aigbavboa, C. (2016). Evaluating barriers to effective implementation of project monitoring and evaluation in the Ghanaian construction industry. *Creative Construction Conference 2016 (CCC 2016), 25–28* June 2016, *Procedia Engineering*, 164 (2016), 389–394, Available online at www.sciencedirect.com

Tengan, C. & Aigbavboa, C. (2018). The role of monitoring and evaluation in construction project management. In: W. Karwowski and T. Ahram (eds.). *Intelligent Human Systems Integration, Advances in Intelligent Systems and Computing*, 722, 571–582. California, USA: Springer. https://doi.org/10.1007/978-3-319-73888-8_89

Vee, C. & Skitmore, M. (2003). Professional ethics in the construction industry. *Engineering, Construction and Architectural Management*, 10(2), 117–127. doi:10.1108/09699980310466596

Zhao, L., Mbachu, J. & Domingo, N. (2017). *Exploratory factors influencing building development costs in New Zealand using SEM.* doi: 10.20944/preprints201704.0187.v1.

2 Overview of project monitoring and evaluation research

2.1 Abstract

The concept of monitoring and evaluation (M&E) has become a vital tool for effective construction project delivery. Undoubtedly, M&E has been hypothesized differently across various fields such as health and agriculture. Due to the varying understanding and conceptualization and different levels of perceptions regarding construction M&E, this chapter provides a theoretical understanding of the concept of M&E in the construction context, i.e. the chapter established that the need for M&E cannot be overemphasized in the achievement of successful project delivery. This understanding is, thus, central to establishing strategies that may help project teams effectively and ultimately undertake M&E to improve the project performance and management.

2.2 Introduction

Extant studies across various domains have underscored the imperative role of M&E in achieving project success (Chin, 2012; Ika, Diallo & Thuillier, 2012; Kamau & Mohamed, 2015; Otieno, 2000; Papke-Shields, Beise & Quan, 2010; Tache, 2011). However, from a theoretical perspective, the term M&E has been conceptualized and defined variedly. The lack of comparable definition of M&E remains a crucial challenge in the literature (Patton, 2003). In the light of the above, this chapter provides an understanding of the concept of M&E as well as the complementary role and difference between M&E, the approach, methods, tools and techniques of M&E and, finally, the benefits and challenges of the M&E practice are placed in perspective. To obtain a better understanding of the diverse and complementary nature of M&E in project delivery, this chapter describes M&E separately and presents the distinction between the two management functions as they influence project success.

2.2.1 Monitoring

Monitoring is a continuous management function that logically collects data on specific indicators of the project to offer management and stakeholders of an ongoing intervention the indications of the extent to which objectives are being achieved and the progress in the use of allocated resources (Omonyo, 2015). Whereas Gudda (2011) described monitoring as an art of collecting project information to make an informed decision at the right time, Ile, Eresia-Eke and Allen-Ile (2012) defined monitoring as an ongoing process of generating information about the progress being made towards the achievement of results. According to Bamberger and Hewitt (1986), monitoring is an internal project activity designed to provide constant feedback on progress or otherwise and the efficiency with which it is being implemented. Monitoring is undertaken while a project is being implemented with the aim of improving the project design and functioning while in action. Otieno (2000) emphasizes that monitoring assesses the understanding of stakeholders regarding the project and promotes the systematic and professional management, minimizes the risk of project failure and, finally, assesses progress in implementation. The World Bank, cited by Tache (2011), agreed in context with the several descriptions of monitoring and further defines monitoring as a continuous assessment of project implementation regarding agreed schedule and use of inputs, infrastructure and services by the beneficiaries. Chipato (2016) identified monitoring as the ongoing process by which stakeholders obtain regular feedback on the progress being made towards achieving their goals and objectives.

Chipato (2016) further stressed that monitoring implies a continuing operation conducted during project implementation to ensure that the project stays on track to achieve its objectives. Also, the International Federation of Red Cross and Red Crescent Societies (IFRC) (2011) defined monitoring as the routine collection and analysis of information to track progress against set plans and check compliance to established standards. It helps identify trends and patterns, adapts strategies and informs decisions for project/programme management. The monitoring exercise may be used to improve project efficacy during implementation: the project should be flexible and able to change and adapt to conditions on the ground as indicated by the exercise (Chipato, 2016). Monitoring is conducted after a programme has begun and continued throughout the programme implementation period. Monitoring is sometimes referred to as process, performance or formative evaluation (Kusek & Rist, 2004).

2.2.1.1 Types of monitoring

In monitoring studies, many forms of monitoring have been identified. Tache (2011) categorizes monitoring into three main types. They are baseline monitoring, impact monitoring and compliance monitoring. Baseline monitoring refers to the measure of economic, social and environmental needs at the pre-project stage to determine the existing situation, range of differences and the process of change, while impact monitoring is the quantification of social and environmental variables during project development and operations to determine the impact that may have been caused by the project (Tache, 2011). Compliance monitoring concerns are ensuring compliance with donor regulations, contract requirements, local building regulations and bye-laws as well as ethical standards in the delivery of the project (Tache, 2011).

Kusek and Rist (2004) categorize monitoring into activity-based monitoring, results-based monitoring and implementation monitoring. Activity monitoring, also known as process monitoring, focuses on how activities are implemented to meet time and cost. However, it falls short in aligning the activities to the outcomes which makes it difficult to understand how these activities have triggered the achievement of the improved performance (Kusek & Rist, 2004). Activity monitoring also tracks the utilization of inputs and resources, the progress of activities and the delivery of outputs. It observes how activities are delivered – the efficiency in time and resources (IFRC, 2011). Results-based monitoring, on the other hand, is explained by Kusek and Rist (2004) to relate to the monitoring of the overall goal of a project and how it affects society. This type of monitoring merges M&E to access whether projects are on target to achieve results (IFRC, 2011). Results-based monitoring is recognized as broad-based and brings activities, processes, inputs and outputs in line with outcomes. Results-based monitoring has been christened as the ideal form of monitoring. Kusek and Rist (2004) further describe implementation monitoring as one which concerns tracking the approaches towards achieving given outcomes in the delivery of projects. This ensures that the right inputs and activities are used to generate output and that there is compliance to achieve set outcomes. Monitoring, therefore, concentrates on resources, activities, objectives (results, purpose development goal) and any fundamental assumptions.

In 2011, the IFRC provided a summary of the different types of monitoring commonly found in a project/programme monitoring system as described in Table 2.1. It is important to remember that these monitoring types often co-occur as part of an overall monitoring system (IFRC, 2011).

2.3 Evaluation

The term "evaluation" has been in the English language for centuries and it has had diverse functions and meanings during that time. Only in recent decades and, in particular, the latter part of the twentieth century, has more precision been given to the term, including specificity to the basic concepts and more explicit explanations about its aims as a functioning entity (Stufflebeam & Shinkfield, 2007). Stufflebeam and Shinkfield (2007) have it that one of the earliest and still most prominent definitions states that it means determining whether objectives have been achieved. Another prominent and widely accepted definition by the Joint Committee set up in 1975, comprising members from 15 professional societies in the United States and Canada aimed at improving evaluation in education that defined "…evaluation as the systematic assessment of the worth or merit of an object" (Stufflebeam & Coryn, 2014). Evaluation arguably is society's most fundamental discipline as it is oriented towards assessing and helping in the improvement of the society at large (Stufflebeam & Shinkfield, 2007). In this context, the construction society comprises the professionals involved in the day-to-day activities on-site and off-site, the equipment and tools used and the products of the industry. It permeates all areas of scholarship, production and service and

Table 2.1 Common types of monitoring

Results monitoring	Results monitoring merges with an evaluation to determine whether the project/programme is on target towards its intended results (outputs, outcomes, impact) and whether there may be any unintended impact (positive or negative). For example, a psychosocial project may monitor that its community activities achieve the outputs that contribute to community resilience and ability to recover from a disaster.
Process (activity) monitoring	Process (activity) monitoring tracks the use of inputs and resources, the progress of activities and the delivery of outputs. It examines how activities are delivered – the efficiency in time and resources. It is often conducted in conjunction with compliance monitoring and feeds into the evaluation of impact. For example, a water and sanitation project may monitor that targeted households receive septic systems according to schedule.
Compliance monitoring	Compliance monitoring ensures compliance with donor regulations and expected results, grant and contract requirements, local governmental regulations and laws and ethical standards. For example, a shelter project may monitor that shelters adhere to agreed national and international safety standards in construction.
Context (situation) monitoring	Context (situation) monitoring tracks the setting in which the project/ programme operates, especially as it affects identified risks and assumptions, but also any unexpected considerations that may arise. It includes the field as well as the larger political, institutional, funding and policy context that affect the project/programme. For example, a project in a conflict-prone area may monitor potential fighting that could not only affect project success but endanger project staff and volunteers.
Beneficiary monitoring	Beneficiary monitoring tracks beneficiary perceptions of a project/ programme. It includes beneficiary satisfaction or complaints with the project/programme, including their participation, treatment, access to resources and their overall experience of change. Sometimes referred to as beneficiary contact monitoring (BCM), it often includes stakeholder complaints and feedback mechanisms. It should take account of different population groups as well as the perceptions of indirect beneficiaries (e.g. community members not directly receiving a good or service). For example, a cash-for-work programme assisting community members after a natural disaster may monitor how they feel about the selection of programme participants, the payment of participants and the contribution the programme is making to the community (e.g. are these equitable?).
Financial monitoring	Financial monitoring accounts for costs by input and activity within predefined categories of expenditure. It is often conducted in conjunction with compliance and process monitoring. For example, a livelihood project implementing a series of micro-enterprises may monitor the money awarded and repaid and ensure implementation is according to the budget and time frame.
Organizational monitoring	Organizational monitoring tracks the sustainability, institutional development and capacity building in the project/programme and with its partners. It is often done in conjunction with the monitoring processes of the larger, implementing organization. For example, a national society's headquarters may use organizational monitoring to track communication and collaboration in project implementation among its branches and chapters.

Source: International Federation of Red Cross, 2011

has important implications for maintaining and improving services and protecting citizens in all areas of interest to society (Stufflebeam & Shinkfield, 2007).

According to Tache (2011), evaluation is the objective and systematic assessment of project activities to determine its relevance, effectiveness, efficiency and impact. It looks at the extent to which objectives have been met, drawing on the data and information generated through monitoring (Otieno, 2000). Project evaluation plays a significant role in the implementation of projects; it informs the decision-making process for improvement, ensures organizational learning from experience to help improve future monitoring and evaluation practice and managers of projects also acquire new skills to be better managers in future (Njama, 2015). Evaluation also helps organizations know their strengths, weaknesses, opportunities and threats (SWOT) for effective M&E (Calder, 2013; Spaulding, 2014). Further, evaluation establishes benchmarks to guide the evaluation of other projects through the creation of a knowledge bank for management, thus organizational learning (Calder, 2013). Evaluation provides the basis for concluding on the efficiency, effectiveness and success or failure of projects (Spaulding, 2014).

An evaluation is a systematic investigation of the value of a programme or other evaluand (the object of an evaluation) or evaluee in the case of a person, which could be a project (Chipato, 2016). Stufflebeam and Shinkfield (2007) opined that evaluands might be individuals, programmes, projects, policies, products, equipment, services, concepts and theories or organizations. More specifically, an evaluation is a process of delineating, obtaining, reporting and applying descriptive and judgmental information about some object's merit, worth, probity and significance (Funnell & Rogers, 2011). Evaluations may involve multiple values of individuals, organizations or societies and these may compete (Stufflebeam & Shinkfield, 2007). Stufflebeam and Shinkfield (2007) further explained evaluation to be a process for giving attestations on such matters as reliability, effectiveness, cost-effectiveness, efficiency, safety, ease of use and probity. Stufflebeam and Shinkfield (2007) also submitted that evaluation serves society by providing affirmations of worth, value, improvement (stating when and how it should occur), endorsement, accountability and, when necessary, a basis for terminating bad programmes. In carrying out an evaluation, the evaluator needs to pay attention to evaluation's root term value in addressing the merit, the worth, the probity and the significance regarding excellence, utility, uncompromising adherence to the basic moral standards, reach, importance and visibility of the project (Frankel & Gage, 2007).

The scope of the evaluation is very wide; it could be personnel evaluation, product evaluation, institutional evaluation, student evaluation and policy evaluation depending on the reason for the evaluation and the variables involved at the particular moment of the evaluation (Frankel & Gage, 2007). The scope of evaluation applications broadens significantly when one considers the wide range of disciplines to which it applies (Stufflebeam & Shinkfield, 2007). One can speak of evaluation, according to Stufflebeam and Shinkfield (2007), to include educational evaluation, social services evaluation, art evaluation, city planning and evaluation, real estate appraising, engineering testing and evaluation, hospital evaluation, drug testing, manufacturing evaluation, consumer products evaluation, agricultural experimentation and environmental evaluation.

2.3.1 Types of evaluation

Evaluation can be classified into several types depending on the intended purpose or stage of its implementation (Omonyo, 2015). Considering the timing of undertake evaluation, three types of evaluation exist according to Tache (2011) and Gudda (2011). They are ex-ante evaluation which is the evaluation conducted before the implementation of the project and in-vivo or mid-term evaluation, also known as interim evaluation, which is undertaken during the implementation of the project. The mid-term evaluation is formative in nature and can occur several times depending on the information needed while the final evaluation describes the evaluation which is conducted after project implementation. The final evaluation is summative in nature and is often conducted by external evaluators. The ex-post evaluation, also described as an evaluation conducted after the period of the final evaluation, is done purposely to ascertain the level of sustainability or impact of the project on the beneficiary community (IFRC, 2011). Other classifications are considered based on the implementers or those responsible for the evaluation, namely whether evaluation is undertaken internally by persons directly involved with the project or externally by stakeholders and donor agencies (Igbokwe-Ibeto, 2012). The IFRC (2011) further identifies participatory evaluation as one which is conducted with the involvement of beneficiaries and all other key stakeholders with the aim to empower, build capacity, ownership and support for the project.

Joint evaluation is also conducted in collaboration with more than one implementing partner, which can help build consensus at different levels as well as credibility and joint support. Joint evaluation has shown numerous benefits which include strengthening evaluation through harmonization and capacity development; shared good practice, innovations and improved programming; reduced transaction costs and management burden (mainly for the partner country); improved donor coordination and alignment; increased donor understanding of government strategies, priorities and procedures; and greater learning by providing opportunities for bringing together wider stakeholders. Learning from evaluation then becomes broader than simply organizational learning: it also encompasses advancement of knowledge in development (UNDP, 2009). The IFRC (2011) brings to light the objectivity and legitimacy of joint evaluation. This enables a greater diversity of perspectives and consensus building, broader scope, being able to tackle more complex and wider-reaching subject areas and enhanced ownership through greater participation (IFRC, 2011).

Evaluation is also viewed based on its focus, whether for accountability purposes (summative) or learning and improvement of performance by management (formative). Also, the IFRC (2011) distinguishes the summative and formative types of evaluation based on the time they occur or whether they are undertaken within the project implementation process. Formative evaluation occurs during project implementation to improve performance and assess compliance, whereas summative evaluation occurs at the end of project/programme implementation to evaluate effectiveness and impact (ibid).

The IFRC (2011) subsequently describes five types of evaluation based on the methodology adopted. These include a real-time evaluation which concerns evaluation during project implementation to provide immediate feedback for modifications to improve ongoing implementation with emphasis on direct lessons learnt and meta-evaluation which describes the type of evaluation that is used to assess the evaluation process itself. The others are thematic evaluation focusing on one theme such as quality, cost or time and is usually undertaken across several projects. Cluster or sector evaluation also focuses on a set of related activities of a project across multiple sites and, finally, impact evaluation which focuses on the effect of a project, rather than on its management and delivery (ibid).

2.3.2 Need for evaluation

M&E of development activities provide government officials, development managers and civil society with better means of learning from experience, improving service delivery, planning and allocating resources and demonstrating results as part of accountability to key stakeholders. Within the development community, there is a strong focus on results and this helps explain the growing interest in M&E. The presence of sound evaluation does not necessarily guarantee high quality in services or that those in authority will heed the lessons of evaluation and take the needed corrective actions but provide only one of the ingredients needed for quality assurance and improvement (Stufflebeam, Madaus & Kellaghan, 2000). There are many examples of defective products that have harmed consumers not because of a lack of pertinent evaluative information, but because of a failure on the part of decision makers to heed and act on rather than ignore or cover up alarming evaluation information (Stufflebeam et al., 2000). One clear example was the continued sales of the Corvair automobile after its developers and marketers knew of its rear-end collision fire hazard. Here we see that society has a critical need not only for competent evaluators but evaluation-oriented decision makers as well (Stufflebeam & Shinkfield, 2007).

For evaluations to make a positive difference, policy-makers, regulatory bodies, service providers and others must obtain and act responsibly on evaluation findings. Stufflebeam and Shinkfield (2007) believed that everyone who plays a decision-making role in serving the public should obtain and act responsibly on evaluations of their services. Fulfilling this role requires each such decision maker to take appropriate steps to become an effective, conscientious, evaluation-oriented service provider. The production and appropriate use of sound evaluation are one of the most vital contributors to strong services and societal progress (Stufflebeam & Shinkfield, 2007).

2.4 Monitoring and evaluation

In juxtaposing M&E definitions, it is evident that they have distinct, yet complementary, functions and roles to play in the life cycle of project delivery. Monitoring gives information on progress of work at any given time (and over time) relative to the planned or desired targets and outcomes, which is descriptive in intent (Kusek & Rist, 2004). Evaluation, on the other hand, gives evidence of the extent to which targets and outcomes are being achieved and it mainly seeks to address issues of causality. Kusek and Rist (2004) further explained that evaluation is a complement to monitoring in that when a monitoring system sends signals that the efforts are going off track (for example, if progress towards the target is lagging and that project duration will not be achieved), then good evaluative information can help address the major issues causing the delays.

In project management studies, several types of research refer to monitoring and control rather than M&E (PMI, 2013). However, it considers the similar description of their roles and functions in project management as a process of tracking, reviewing and regulating progress to achieve performance objectives (Kamau & Mohamed, 2015). M&E are also being seen in an economic sense as a study of human behavior in the process of using resources to attain developmental goals.

The combined effort of both M&E seeks to guarantee efficiency and effective utilization of resources and processes in the project life cycle towards the achievement of successful project delivery. M&E have also been viewed in two main perspectives, that is, the classical/traditional view and the modern view (Tache, 2011). Tache (2011) further describes the traditional (classical) view to acknowledge a distinctive nature of the two management functions, whereas the modern view of M&E is a closely intimate management function. Further, the classical view of M&E limits monitoring to the collection of information on input and output with evaluation taking place once or twice during the implementation of the project. Monitoring and evaluation also focused on assessing the efficiency of projects and focused mostly on project objectives and budget (Tache, 2011). Nonetheless, the classical or traditional approach did not afford stakeholders and project managers the opportunity to appreciate the underlying reasons for the success or failure of the project and this is the major strength of the modern approach to monitoring and evaluation (Kusek & Rist, 2004).

On the other hand, the modern view of M&E acknowledges M&E as an integrated management function (Tache, 2011). This means that the monitoring function will not be complete in measuring performance until the monitored information has been evaluated to establish the extent of success or failure during project implementation. Also, the modern view of M&E considers a broader view of the practice and the approach to M&E is closely related. M&E are seen to focus on all components of the project such as activities, budget, results, risk and objectives (Tache, 2011). Mwangu and Iravo (2015) recognizes the distinct functions of M&E while admitting that their relationship cannot be overemphasized.

Scholars like Cameron (1993) indicate two reasons why M&E are regarded as two distinct functions. These he informs as the failure of M&E to come up with useful and cost-effective information regarding decision making due to the lack of trust in collecting accurate data and failure to process and analyze the same. Secondly, M&E were concerned with responding to planning failures rather

Table 2.2 Complementary roles of M&E

Monitoring	Evaluation
• Clarifies programme objectives	• Analyzes why intended results were or were not achieved
• Links activities and their resources to objectives	• Assesses specific casual contributions of activities to results
• Translates objectives into performance indicators and sets targets	• Examines the implementation process
• Routinely collects data on these indicators, compares actual results with targets	• Explores unintended results
• Reports progress to managers and alerts them to the problems	• Provides lessons, highlights significant accomplishments or programme potentials and offers recommendations for improvement

Source: Kusek & Rist (2004)

than the conventional managerial ineffectiveness and inefficient implementation (Cameron, 1993). Hence, M&E is part of a similar process which is being implemented through all the phases in the project life cycle and covers all the knowledge areas recognized in the Project Management Body of Knowledge (PMBOK) published by the Project Management Institute (PMI, 2013). M&E functions have both distinct and complementary roles in project implementation. Kusek and Rist (2004) outline some major complementary roles of monitoring and evaluation in Table 2.2.

M&E as having been acknowledged in this study is undertaken throughout the entire life cycle of the project. Various forms of M&E can, therefore, be identified based on the time and focus of assessment (Kusek & Rist, 2004). Whereas monitoring tends to focus on what is happening continuously, evaluations, on the other hand, are conducted at a specific time to assess how successful the project is (progress against programme) and what difference or impact the project has made (Otieno, 2000). Monitoring data is typically used by managers for ongoing project implementation, tracking outputs, budgets and compliance with procedures. Evaluations may also inform implementation (e.g. a mid-term evaluation), but they are less frequent and examine more significant changes (outcomes) that require more methodological rigour in analysis such as the impact and relevance of an intervention (IFRC, 2011). In furtherance of the stated differences that exist between M&E as discussed, Table 2.3 presents a summary of the differences between M&E. The identified difference between M&E is based on some important questions to explain the significant differences between the two management functions.

2.5 Approach, methods, tools and techniques of M&E

Several studies refer to the same construct with different terminologies. This could be attributed to the contextual use of the terminologies. For instance, the terms "approach", "methods", "tools" and "techniques" for M&E have been used

Table 2.3 Differences between M&E

Question	Monitoring	Evaluation
When is it done?	• Continuously – throughout the life of the project/programme	• Occasionally – before implementation, mid-term, at the end or beyond the project/ programme period
What is measured?	• Efficiency – use of inputs, activities, outputs and assumptions	• Effectiveness, longer-term impact and sustainability – achievement of purpose and goal and unplanned change
Who is involved?	• Staff within the agency (internal staff)	• In most cases, external bodies or agencies are engaged in the evaluation
Sources of Information	• Internal documents, e.g. monthly or quarterly reports, work and travel logs, minutes of meetings	• Internal and external documents, e.g. consultant's reports, annual reports and national statistics
Who uses the results?	• Managers and project/ programme staff	• Managers, staff, funding agency (e.g. CDC) beneficiaries and other agencies
How are results used?	• To make minor changes	• To make major changes in policy, strategy and future work

Source: Stufflebeam, 2003

differently to mean the same constructs. M&E are regarded as a performance measurement or assessment tool undertaken to ensure the achievement of set standards such as quality, cost or time. It may also seek to measure the impact (functionality) of the construction project on beneficiary communities (impact measurement), thus measuring project performance (Elazouni & Salem, 2011). It is, therefore, necessary for the right M&E approach, method, tools and techniques to be employed to generate the right progress data for comparison with the existing success indicators for decision making on project delivery.

Barasa (2014) argues that the contribution of M&E tools to the success of projects if rightly used is enormous. Since projects are unique and revolving, Otieno (2000) informs that several measurement tools have been developed to measure various indicators over the life of the project. Specific project performance indicators include quality, cost, time/schedule, performance, satisfaction, health and safety. In most cases, different methods, tools and techniques are used in the M&E for the purpose of achieving different objective targets of the project. M&E data may be gathered in multiple ways: firstly, through the traditional M&E where M&E teams visit project sites and manually record project progress information to evaluate progress made on the project. This method, however, could slow down the M&E process causing delay to the entire project as well as compromising on the quality of data collected because of the great variation in the reporting skills as well as the willingness to record data accurately (Sacks, Navon & Goldschmidt, 2003).

According to Kumaraswamy (1991) and Pramod, Phaniraj and Srinivasan (2014), a number of scheduling and progress control techniques have been identified such as bar charts, earned value method/analysis, the critical path method (CPM) and network float, whereas some limitations to provide spatial information on construction projects were noticed (Poku & Arditi, 2006). Al-Jibouri (2003) studied the effectiveness of monitoring systems on project cost control in construction and discussed the utility of the leading parameter technique, variance method and activity-based ratios technique in measuring the extent of the project cost at stages of the project.

Compared to the unit cost technique, the leading parameter technique considers the use of one or more major types of work packages to measure the overall performance of the project. For instance, it can be used as a measure of the cost performance of the entire project in the case of a cost-significant item such as concrete, and reinforcement during construction. A comparison of the cost of the leading parameter work package and the overall project cost is made at the same stage of M&E to establish the percentage of progress. The activity-based ratio is a project cost-control technique that compares the ratios of incomes and expenditures on project activities to measure the performance of the entire project. Three ratios are identified, namely a planned performance which is measured as a ratio of the expected incomes to the planned expenditures; actual performance which is the ratio of actual incomes to the actual expenditures; and efficiency ratio which measures the ratio of actual performance to planned performance (Al-Jibouri, 2003). Finally, the variance and earned value analysis methods measure project performance by comparing the current and final stages of the actual and planned expenditures of the project. This method assesses the entire project or sections of the project integrating cost and time (Khamidi, Khan & Idrus, 2011).

In the M&E of project quality, two sets of documents are used to determine the level of quality of the activity or work. They are the specification and the drawings (Harris, McCaffer & Edum-Fotwe, 2013). Further, Harris et. al. (2013) postulate that the main quality M&E techniques are through field or site inspections and statistical techniques which are based on the sampling of construction products to ascertain the level of quality against specified requirements. The former is fundamentally subjective and difficult to undertake. The statistical technique collects samples in the form of concrete cubes or masonry blocks which are sent to the laboratory for strength and other property testing.

The data collected with the above performance measurement techniques could be fed into computer systems to produce some meaningful results for management decision to be made. Data collection is automated and produces immediate and more accurate data on project progress for management use. Several computer-aided data-gathering tools and techniques have been studied for viewing and monitoring the progress of construction projects (Elazouni & Salem, 2011). Elazouni and Salem (2011) proposed a pattern recognition (PR) concept and technique based on the technique of critical path method (CPM) to monitor and evaluate the overall progress of the projects. Automating the entire construction process to ensure effective performance M&E necessitated the advent of computer-aided design (CAD) and the building information modelling (BIM) tools. The CAD and BIM as monitoring and evaluation tools helped generate and maintain construction schedules from architectural drawings for constructability review and project planning (Yadhukrishnan & Shetty, 2015).

Owing to the limitation of the CPM and bar charts, system integration of a progress monitoring system (PMS) and geographical information system (GIS) was developed to represent both construction progress in the form of a CPM schedule as well as a graphical representation of synchronized construction work schedule (Poku & Arditi, 2006). Before the work of Poku and Arditi (2006), Yeh and Li (1997) studied the use of GISs and remote sensing techniques and described them as valuable tools in the formulation, implementation and monitoring of urban development projects in the move towards sustainable project development. Integration of GIS and database management system for scheduled monitoring was reported by Cheng and Chen (2002) for precast building construction. Li, Chen, Yong and Kong (2005) described the application of an integrated global positioning system (GPS) and GIS in reducing construction waste.

Also, in monitoring the location of construction equipment to ensure safety and purposeful usage, Sacks, Navon, Brodetskaia and Shapira (2005) reported on the use of a "black box" monitor and an electronic building information model. Also, with the help of a computerized building project model (BPM), integrated with the physical geometry of the building, the monitoring of labour inputs was achieved (Sacks, et al., 2003). Project performance measures were obtained from the comparison of the most reasonable budget and duration values from individual probability distributions for actual progress, with the project's actual data and cumulative cost while stressing the unavailability of relevant historical data as a major limitation (Elazouni et al., 2011). Similarly, Nassar, Gunnarsson and Hegab (2005) reported on the Weibull analysis used in combination with the earned value method (EVM) to evaluate the scheduled performance of construction projects. The stochastic S-curve also aided in estimating the probable expected cost and duration of projects (Barraza, Back & Mata, 2000). The Project Management Institute (PMI, 2013) also identifies expert judgement; analytical techniques such as the regression analysis, grouping methods, causal analysis and root cause analysis; forecasting methods (e.g. time series, scenario building, simulation), failure mode and effect analysis (FMEA); fault tree analysis (FTA); reserve analysis; trend analysis; earned value management; variance analysis; project monitoring and information systems; and meetings as tools and techniques used in monitoring and evaluating projects.

2.6 Monitoring and evaluation indicators

Often M&E are characterized by comparing performance with set standards (Stufflebeam & Shinkfield, 2007). This perhaps underscores why standards such as quality, time, cost and satisfaction levels of projects are set as targets before projects commence. Ile et al. (2012) describe indicators as a measurement tool that aids project managers to ascertain the extent to which results are being achieved.

Prennushi, Rubio and Subbarao (2001) also define indicators as variables used to measure progress towards the goals. An indicator is a sign which specifies the progress of project intervention, whether project objectives are being met (Rugg, 2010). Hammond, Adriaanse, Rodenburg, Bryant and Woodward (1995) posit that indicators communicate information about progress towards achieving impact and provide clues about matters of greater implication or a phenomenon that is not immediately apparent. The impact is the positive or negative, planned or unplanned effect occasioned by a development project. Measuring the impact of a project, therefore, is critical in producing valuable information for the decision-making process and supports accountability for the delivery of results (Njama, 2015). Kusek and Rist (2004) emphasize that performance targeting of indicators and assessment of progress towards its achievement provide early warnings to allow corrective measures to be taken, indicating whether an in-depth evaluation or review is required.

Indicators provide both qualitative and quantitative data which offers a simple and consistent approach to monitor, measure and determine performance and achieve accountability (Kusek & Rist, 2004; Ile et al., 2012). Also, performance indicators present an effective means to measure progress towards objectives and facilitate benchmarking comparisons between different organizational units (Mosse & Sontheimer, 1996). Indicators are developed for all levels of the M&E system; thus, indicators are imperative for monitoring progress concerning inputs, activities, outputs, outcomes and impact. Progress needs to be monitored at all levels of the system to provide feedback on areas of achievement and areas in which improvement may be necessary (Kusek & Rist, 2004). To ensure that performance indicators are set right to serve their purpose effectively, several criteria are identified in the literature. The SMART criteria acknowledge that indicators should be Specific, Measurable, Attainable, Results-oriented and Time-bound (Dale, 2003; Larson & Williams, 2009; Mulandi, 2013). Other standards or criteria for determining a good indicator are the SPICED criteria (Roche, 1999). He argues that indicators must be Subjective, Participatory, Interpreted and Communicable, Cross-checked and compared, Empowering and Diverse and disaggregated. Schiavo-Campo (1999), Gudda (2011) and Ile et al. (2012) also outline the CREAM criteria. This implies that all indicators should be Clear, Relevant, Economic, Adequate and Monitorable. Gage and Dunn (2009) also define six characteristics of a good indicator as Valid, Reliable, Measurable, Precise, Timely and Programmatically relevant. It can, therefore, be concluded that indicators must pass the tests of reliability, feasibility and utility in decision making.

2.7 Types of monitoring and evaluation indicators

Two broad types of indicators are identified in the literature. They are qualitative indicators, also known as outcome or performance indicators and quantitative indicators, otherwise referred to as output indicators. Quantitative indicators tell whether the activities are taking place as planned and on the path to success. They do not, however, provide any information on the effect or impact they bring about. Qualitative indicators, on the other hand, are usually concerned with change (outcome). They provide information on changes caused by the project in the lives of beneficiaries. Unlike quantitative indicators which are in numeric forms, qualitative indicators are non-numeric and help determine the level of progress towards the achievement of objectives. It is, therefore, necessary to monitor both process and impact of a project. Omonyo (2015) posits that different indicators must be developed for each level of results, at least one indicator for each level of core activity. This proposition, therefore, suggests a classification of indicators into input indicators, process or activity indicators, output indicators, results or outcome indicators and impact indicators. Input indicators will examine the needs of the project such as the resource requirement to undertake effective M&E while process indicators seek to ensure that plan activities are being done right at every stage of the M&E. Output indicators are relevant to measure the immediate achievement of a project or programme.

2.8 Benefits of effective monitoring and evaluation

Studies have shown a plethora of benefits derived from the effective M&E of projects (Kamau & Mohamed, 2015; Otieno, 2000; Tache, 2011). Implementation of monitoring and evaluation seeks to guarantee ultimate project success through the achievement of immediate project outcomes such as conformity to standards and the achievement of budget and schedule as well as long-term objectives such as fit for purpose (impact). The collective achievement of all immediate outcomes indicates that M&E are effective and, therefore, the success of the project is achieved (Chin, 2012; Ika et al., 2012; Papke-Shields et al., 2010). A study by Papke-Shields et al. (2010) revealed that conformity to project specification (quality) would be achieved when projects are effectively monitored and evaluated. The study further accentuates the achievement of projects within the approved budget (cost) and project duration (time) when M&E are effectively undertaken (Papke-Shields et al., 2010). Further, human organizational capacity and that of stakeholders are developed through effective M&E along with effective communication (Papke-Shields et al., 2010).

Beyond achieving direct project objectives such as cost, time and quality, organizations are afforded with the opportunity to learn (organizational learning) from previous practices and activities to help improve current and future project implementation and better decision making (Chipato, 2016). Donor agencies and project financiers are satisfied with the accountability level of projects given an effective M&E practice ensuring future interest in funding development projects by donors. Contractors are guided through the project implementation process which guarantees the utmost performance of contractors. An effective project M&E practice ensures a healthy project implementation environment where all stakeholders are well represented on the project and given the opportunity to contribute to the project. Also, scarce project resources are committed to judicious use. A greater benefit of effective M&E is the assurance that project activities are done right the first time to eliminate rework (which is a likely contributor to increased project budget and extended project duration) arising from design and construction errors.

M&E activities improve communication between different stakeholders. This affords stakeholders better understanding of implementation issues regarding all aspects of the project. To make communication effective, a constructive environment for exchange and discussion is essential. Clear and transparent communication mechanisms such as regular meetings, workshops, reporting and information sharing via the Internet or printed media should also be established. It can therefore be concluded that the indicators of a successful project such as achieved project time, conformity to standards, achieving project cost, stakeholder satisfaction, contractor performance, health and safety, value for money, environmental performance, end-user satisfaction, client satisfaction and fitness for purpose are achieved through the effective implementation of M&E of projects.

2.9 Challenges to monitoring and evaluation

M&E implementation is challenged on many fronts. This section discusses M&E challenges under three broad categories, namely **technical-level**, **organizational-level** and **project-level** challenges.

2.9.1 *Organizational-level challenges*

At the organizational level, Cameron (1993) discusses the lack of M&E units within the organization as a significant challenge for the M&E of projects. Without an M&E unit, planning responsibilities toward the M&E of projects are rendered ineffective. The need to strengthen the planning and implementation of M&E for efficient project delivery is paramount. When they studied the barriers to M&E implementation in the Ghanaian construction industry, Tengan and Aigbavboa (2016) identified weak institutional capacity as influencing the performance of M&E significantly, as well as lack of technical capacity, skills and knowledge among M&E staff for the M&E process. This resulted in project failure from the onset. In addition, special skills and knowledge are required in planning and undertaking M&E, hence the need for concerted continuous training on M&E for M&E team and project staff. Badom (2016) also laments the seeming non-existence of M&E plans or system integration during planning, budgeting and infrastructure development.

This potentially puts M&E in disarray; there are no set indicators to measure progress and impact and, as such, any level of performance is acceptable. Muriithi and Crawford (2003) further advanced that one major challenge observed at the managerial level of organizations that impact the M&E of the project is the struggle for power between M&E unit staff. In addition, the general organizational structure is said to influence project M&E. This does not allow the M&E unit the independence and self-sufficiency to deal with all setbacks in the M&E of projects. A major attribute of effective M&E is its support for decision making and organizational learning. Unfortunately, M&E information and reports are poorly utilized to inform the organizational planning process and implementation of future projects in the Ghanaian construction industry (Tengan & Aigbavboa, 2016). Significant among the challenges to the adoption of the BIM technology in the UK are the lack of investment and poor demand for its use by clients (Kim & Park, 2013).

2.9.2 Project-based challenges

The successful implementation of M&E at the project level is hinged on the effective planning for M&E at the management level. At the project level, limited financial resources affect the M&E process negatively (Badom, 2016; Cameron, 1993; Tengan & Aigbavboa, 2016). The approaches adopted in collecting project information for decision making renders the quality of data collected poor and inadequate for management to base their decision on for future projects (Tengan & Aigbavboa, 2016). Similarly, the challenges in collecting and analyzing M&E data have been acknowledged which requires the need for relevant information to be generated through effective data collection and analysis to ensure good management decision on M&E activities on the project (Otieno, 2000). Communication during the M&E of projects is critical. Diallo and Thuillier (2005) link project success to communication between key stakeholders on the project. Poor communication is further exacerbated by the insufficient information on project design as well as the inconsistency of project information, namely drawings, specifications and bill of quantities available for the M&E.

2.9.3 Technical-based challenges

Technical challenges rendering the ineffectiveness of monitoring and evaluation of projects are reviewed. According to Bamberger, Rao and Woolcock, (2010) and Chaplowe (2008), the weak demand for evaluation utilization poses a challenge towards implementation. As has been noted in previous sections, the diverse focus of M&E, considering the specific needs of the project stakeholders and donors which have generated several operational definitions toward describing what M&E entails, poses the lack of comparable definitions as opined by Patton (2003). This has created different understandings of effective M&E implementation. Also, Auriacombe (2013) asserts that various attempts made to classify evaluation methods were aimed at simplifying the puzzling array of available methods. However, the attempts have instead created confusion regarding the understanding of the evaluation field. The challenge of weak linkage between planning and M&E is reported by Seasons (2003) and the weak legal and institutional frameworks on monitoring and evaluation (Basheka & Byamugisha, 2015) cannot be overlooked. A sound system must exhibit the characteristics which describe the criteria for assessing the quality of an M&E system. The schedule in Table 2.4 summarizes the various categories challenges in M&E implementation.

Table 2.4 Categories of M&E challenges

Source	Technical	Organizational	Project
		Types of barriers	
Seasons (2003)	Weak linkage between planning and monitoring and evaluation		
Basheka & Byamugisha (2015)	Weak legal and institutional frameworks		
Auriacombe (2013)	Methodological issues		
Patton (2003)	Lack of comparable definitions		
Otieno (2000)			Poor approach to data collection and analysis
Bamberger et al. (2010)	Low rate or weak demand for evaluation utilization		
Badom (2016)		The seeming non-existence of monitoring and evaluation plans in planning, budgeting and infrastructure development	Limited financial resources
Tengan & Aigbavboa (2016)		Weak institutional capacity; poor utilization of M&E information and reports	Limited financial resources; quality of data collected
Diallo & Thuillier (2005)			Poor communication; insufficient information on project design as well as the inconsistency of project information; drawings, specifications and bill of quantities
Muriithi & Crawford (2003)		Power struggles between M&E unit or officers and general organizational structure	
Chaplowe (2008)	Low rate or weak demand for evaluation utilization		
Cameron (1993)		Lack of M&E units within the organization	Limited financial resources
Kim & Park (2013)		Poor demand for M&E	Poor or lack of investment in M&E

Source: Literature

Summary

The chapter provided an overview of understanding of M&E research. The chapter established that the need for M&E cannot be overemphasized in the achievement of successful project delivery. An understanding of the types of M&E, need for effective M&E in project delivery and the complementary roles of M&E were outlined. Further, the appropriate methods and approaches, tools and techniques used in M&E were discussed. The next chapter focuses on the existing M&E frameworks and indicators and their essence in M&E practice. The next chapter will also discuss M&E systems and their components and, finally, the main challenges to M&E implementation.

References

Al-Jibouri, S. H. (2003). Monitoring systems and their effectiveness for project cost control in construction. *International Journal of Project Management*, 21(2), pp. 145–154.

Auriacombe, C. (2013). In search of an analytical valuation framework to meet the needs of government. *Journal of Public Administration*, 48(4.1), pp. 715–729.

Badom, L. N. (2016). *Project monitoring and evaluation: A critical factor in budget implementation, infrastucture development and sustainability.* In: Nigerian Institute of Quantity Surveyors' National Workshop with the Theme: "Budgeting and Capital Project Monitoring and Evaluation in an Era of Change" at Precious Conference Centre, Benue Hotels, Makurdi, Benue State, Nigeria on 27th - 28th July, 2016

Bamberger, M., Rao, V. & Woolcock, M. (2010). *Using mixed methods in monitoring and evaluation: experiences from international development.* The University of Manchester: Brooks world poverty Institute, BWPI Working Paper 107

Bamberger, M. & Hewitt, E. (1986). *Monitoring and evaluating urban development programs: A handbook for program managers and researchers.* Washington, DC: World Bank.

Barasa, R. M. (2014). *Influence of monitoring and evaluation tools on project completion in Kenya: A case of Constituency Development Fund projects in Kakamega County, Kenya.* Kenya: University of Nairobi.

Barraza, G. A., Back, W. E. & Mata, F. (2000). Probabilistic monitoring of project performance using SS-Surve. *Journal of Construction Engineering and Management*, 126(2), pp. 142–148.

Basheka, B. C. & Byamugisha, A. (2015). The state of monitoring and evaluation (M&E) as a discipline in Africa. *African Journal of Public Affairs*, 8(3), pp. 75–95.

Calder, J. (2013). *Programme evaluation and quality: A comprehensive guide to setting up an evaluation system.* London: Routledge.

Cameron, J. (1993). The challenges for monitoring and evaluation in the 1990s. *Project Appraisal*, 8(2), pp. 91–96, doi:10.1080/02688867.1993.9726893.

Chaplowe, S. G. (2008). *Monitoring and evaluation planning: Guidelines and tools.* Washington, DC and Baltimore, MD: Catholic Relief Services.

Cheng, M.-Y. & Chen, J.-C. (2002). Integrating barcode and GIS for monitoring construction progress. *Automation in Construction*, 11(1), pp. 23–33.

Chin, C. M. M. (2012). *Development of a project management methodology for use in a university-industry collaborative research environment.* Nottingham: University of Nottingham.

Chipato, N. (2016). *Organisational learning and monitoring and evaluation in project-based organisations.* Stellenbosch: Stellenbosch University.

Dale, R. (2003). The logical framework: An easy escape, a straitjacket, or a useful planning tool? *Development in Practice*, 13(1), pp. 57–70, doi:10.1080/0961452022000 037982.

Diallo, A. & Thuillier, D. (2005). The success of international development projects, trust and communication: an African perspective. *International Journal of Project Management*, 23(3), pp. 237–252, doi:10.1016/j.ijproman.2004.10.002.

Elazouni, A. & Salem, O. A. (2011). Progress monitoring of construction projects using pattern recognition techniques. *Construction Management and Economics*, 29(4), pp. 355–370, doi:10.1080/01446193.2011.554846.

Frankel, N. & Gage, A. (2007). *M&E fundamentals: A self-guided minicourse.* Washington D.C.: U.S. Agency for International Development, MEASURE Evaluation, Interagency Gender Working Group.

Funnell, S. C. & Rogers, P. J. (2011). *Purposeful program theory: Effective use of theories of change and logic models.* John Wiley & Sons.

Gage, A. & Dunn, M. (2009). *Monitoring and evaluating gender-based violence prevention and mitigation programs.* Washington DC. Galtung, J: US Agency for International Development, MEASURE Evaluation, Interagency Gender Working Group.

Gudda, P. (2011). *A guide to project monitoring & evaluation.* United States of America: AuthorHouse.

Hammond, A., Adriaanse, A., Rodenburg, E., Bryant, D. & Woodward, R. (1995). *Environmental indicators: A systematic approach to measuring and reporting on environmental policy performance in the context of sustainable development.* Washington, DC: World Resources Institute.

Harris, F., McCaffer, R. & Edum-Fotwe, F. (2013). Modern Construction Management. Wiley. ISSBN 9781118510186

Igbokwe-Ibeto, C. J. (2012). Issues and challenges in local government project monitoring and evaluation in Nigeria: The way forward. *European Scientific Journal*, 8(18).

Ika, L. A., Diallo, A. & Thuillier, D. (2012). Critical success factors for World Bank projects: An empirical investigation. *International Journal of Project Management*, 30(1), pp. 105–116, doi:10.1016/j.ijproman.2011.03.005.

Ile, I. U., Eresia-Eke, C. & Allen-Ile, C. (2012). *Monitoring and evaluation of policies, programmes and projects.* Hatfield, Pretoria: Van Schaik.

International Federation of Red Cross and Red Crescent Societies (IFRC). (2011). *Project/programme monitoring and evaluation guide.* No. 1000400 E 3,000 08/2011. Geneva: IFRC.

Joint United Nations Programme on HIV/AIDS (UNAIDS) (2009). *12 components monitoring and evaluation system strengthening tool.* Geneva: UNAIDS.

Kamau, C. G. & Mohamed, H. B. (2015). Efficacy of monitoring and evaluation function in achieving project success in Kenya: A conceptual framework. *Science Journal of Business and Management*, 3(3), p. 82, doi:10.11648/j.sjbm.20150303.14

Khamidi, M. F., Khan, W. A. & Idrus, A. (2011). The cost monitoring of construction projects through earned value analysis. *Presented at the IPEDR International Conference on Economics and Finance Research*, Singapore: IACSIT Press, pp. 124–128.

Kim, K. P. & Park, B. L. (2013). BIM feasibility study for housing refurbishment projects in the UK. *Organization, Technology & Management in Construction: An International Journal*, 6(2), pp. 765–774, doi:10.5592/otmcj.2013.2.1.

Kumaraswamy, M. M. (1991). *Evaluating the management of construction projects.*

Kusek, J. Z. & Rist, R. C. (2004). *Ten steps to a results-based monitoring and evaluation system: A handbook for development practitioners.* Washington, DC: World Bank.

Larson, S. & Williams, L. J. (2009). Monitoring the success of stakeholder engagement: Literature review. *People, communities and economies of the Lake Eyre Basin*, pp. 251–298.

Li, H., Chen, Z., Yong, L. & Kong, S. C. W. (2005). Application of integrated GPS and GIS technology for reducing construction waste and improving construction efficiency. *Automation in Construction*, 14(3), pp. 323–331, doi:10.1016/j.autcon.2004.08.007.

Mosse, R. & Sontheimer, L. E. (1996). *Performance monitoring indicators handbook.* Washington, DC.

Mulandi, N. M. (2013). *Factors influencing performance of monitoring and evaluation systems of Non-governmental organizations in governance: A case of Nairobi, Kenya.* University of Nairobi.

Muriithi, N. & Crawford, L. (2003). Approaches to project management in Africa: Implications for international development projects. *International Journal of Project Management*, 21(5), pp. 309–319, doi:10.1016/S0263-7863(02)00048-0.

Mwangu, A. W. & Iravo, M. A. (2015). How monitoring and evaluation affects the outcome of Constituency Development Fund projects in Kenya: A case study of projects in Gatanga Constituency. *International Journal of Academic Research in Business and Social Sciences*, 5(3), doi:10.6007/IJARBSS/v5-i3/1491.

Nassar, K. M., Gunnarsson, H. G. & Hegab, M. Y. (2005). Using Weibull analysis for evaluation of cost and schedule performance. *Journal of Construction Engineering and Management*, 131(12), pp. 1257–1262.

Njama, A. W. (2015). *Determinants of effectiveness of a monitoring and evaluation system for projects: A case of Amref Kenya WASH programme.* Nairobi: University of Nairobi.

Otieno, F. A. O. (2000). The roles of monitoring and evaluation in projects. In: *2nd International Conference on Construction in Developing Countries: Challenges Facing the Construction Industry in Developing Countries*, pp. 15–17.

Omonyo, A. B. (2015). *Lectures in project monitoring & evaluation for professional practitioners.* Germany: Lambert Academic Publishing.

Papke-Shields, K. E., Beise, C. & Quan, J. (2010). Do project managers practice what they preach, and does it matter to project success? *International Journal of Project Management*, 28(7), pp. 650–662, doi:10.1016/j.ijproman.2009.11.002

Patton, M. Q. (2003). Inquiry into appreciative evaluation. *New Directions for Evaluation*, 2003(100), pp. 85–98.

Poku, S. E. & Arditi, D. (2006). Construction scheduling and progress control using geographical information systems. *Journal of Computing in Civil Engineering*, 20(5), pp. 351–360.

Pramod, M., Phaniraj, K. & Srinivasan, V. (2014). Monitoring system for project cost control in construction industry. *International Journal of Engineering Research & Technology*, 3(7), pp. 1487–1491.

Prennushi, G., Rubio, G. & Subbarao, K. (2001). Monitoring and evaluation. *World Bank PRSP Sourcebook*, pp. 105–130.

Project Management Institute (PMI). (2013). *Managing change in organizations: A practice guide.* UK: PMI.

Roche, C. (1999). *Impact assessment for development agencies: Learning to value change.* Oxfam GB.

Rugg, D. (2010). *An introduction to indicators.* UNAIDS Monitoring and Evaluation Fundamentals.

Sacks, R., Navon, R. & Goldschmidt, E. (2003). Building project model support for automated labor monitoring. *Journal of Computing in Civil Engineering*, 17(1), pp. 19–27.

Sacks, R., Navon, R., Brodetskaia, I. & Shapira, A. (2005). Feasibility of automated monitoring of lifting equipment in support of project control. *Journal of Construction Engineering and Management*, 131(5), pp. 604–614.

Seasons, M. (2003). Monitoring and evaluation in municipal planning: Considering the Realities. *Journal of the American Planning Association,* 69(4).

Schiavo-Campo, S. (1999). Strengthening "performance" in public expenditure management. *Asian Review of Public Administration,* 11(2), pp. 23–44.

Spaulding, D. T. (2014). *Program evaluation in practice: Core concepts and examples for discussion and analysis,* 2nd edition. San Francisco, California: Jossey-Bass.

Stufflebeam, D. L. (2003). The CIPP model for evaluation. In: *International handbook of educational evaluation.* Dordrecht: Springer, pp. 31–62.

Stufflebeam, D. L. & Coryn, C. L. S. (2014). *Evaluation theory, models & applications* 2nd edition. San Francisco, CA: Jossey-Bass.

Stufflebeam, D. L., Madaus, G. F. & Kellaghan, T. (Eds). (2000). *Evaluation models: Viewpoints on educational and human services evaluation* 2nd edition. Springer Netherlands.

Stufflebeam, D. L. & Shinkfield, A. J. (2007). *Evaluation theory, models, and applications.* San Francisco, California: Jossey-Bass.

Tache, F. (2011). Developing an integrated monitoring and evaluation flow for sustainable investment projects. *Economia. Seria Management,* 14(2), pp. 380–391.

Tengan, C. & Aigbavboa, C. (2016). Evaluating barriers to effective implementation of project monitoring and evaluation in the Ghanaian construction industry. *Procedia Engineering,* 164, pp. 389–394, doi: 10.1016/j.proeng.2016.11.635.

United Nations Development Programme (UNDP). (2009). *Handbook on planning, monitoring and evaluation for development results.* New York, USA:.UNDP.

Yadhukrishnan, A.V. & Shetty, A. (2015). A review on GIS-based construction project management. *International Advanced Research Journal in Science, Engineering and Technology,* 2(6), pp. 137–141.

Yeh, A. G. & Li, X. (1997). An integrated remote sensing and GIS approach in the monitoring and evaluation of rapid urban growth for sustainable development in the Pearl River Delta, China. *International Planning Studies,* 2(2), pp. 193–210, doi:10.1080/13563479708721678.

3 Monitoring and evaluation system and framework

3.1 Abstract

The knowledge of monitoring and evaluation (M&E) system and framework is critical for effective planning and management of the project life-cycle, providing a clear understanding of the M&E process and the relationship that exist between project activities. Indeed an M&E framework helps enhance understanding, guides the development of an M&E plan, provides the basis to implement M&E activities and helps define the relationship among inputs, activities, output, outcomes and impact. M&E systems guide the implementation process and also present the opportunity to gather relevant results of information regarding progress for effective decision making and efficient communication amongst stakeholders to foster interaction among stakeholders to build team spirit. A detailed description and illustration of key M&E frameworks and systems are discussed in reference to construction project delivery.

3.2 Introduction

In the quest to have a functional M&E system, a clear and working framework is very important to guide the entire process of M&E (Frankel and Gage, 2007). A framework explains how a programme should work by itemizing the components of the initiative and the steps in order to achieve the desired results, i.e. the aims and objectives of the M&E process. The framework further increases the level of understanding of the programme's objectives while giving meaningful definition to the relationship between key factor to implementation and articulates by giving reliable details of the internal and external elements that could affect the programme's success. Ile et al. (2012) define a framework as a skeleton or structure that provides a graphical depiction of the critical components of the project and how elements are systematically interlinked from the beginning to the end. Omonyo (2015) explains a framework as a guide to develop a useful M&E system. Frameworks operationalize how the project is supposed to work by laying out the components and steps required to achieve the desired outcome or results (UNDP, 2009). The framework also outlines the relationship between implementation variables and identifies all other elements that could impede project success.

Shapes such as squares, circles, triangles and stars which are linked with arrows or lines could be used to describe the relationship between different activities in an M&E framework (Ile et al., 2012). The importance of framework can therefore not be underestimated in the study and practice of M&E. Indeed, the lack of an evaluation framework has an adverse effect on project success (Al-Otaibi, 2011). An M&E framework helps enhance understanding, guides the development of an M&E plan, provides the basis to implement M&E activities and helps define the relationship among inputs, activities, output, outcomes and impact (Omonyo, 2015). He further indicates that M&E frameworks describe how activities will lead to the desired output, outcome and impact.

A well-thought-out M&E framework can assist greatly with thinking through programmatic strategies, objectives and planned activities and whether they are indeed the most appropriate ones to implement. Programmes should select the type of framework that best suits their strategies and activities and responds to institutional requirements. An appropriate framework for M&E of activities can be designed and implemented even when programmes do not have significant resources and where a programme staff and implementers, service providers and policy-makers feel they do not have additional time to devote to M&E (Frankel & Gage, 2007). Several M&E frameworks exist with none being superior to the other (Frankel & Gage, 2007). Developing or selecting an M&E framework for a project is founded on the needs or requirements of the stakeholders, clients or donor partners or the specific nature of the project (Ile et al., 2012; Omonyo, 2015). It is important therefore to select appropriate frameworks that will give a clearer picture of the processes in the M&E of projects. This framework, as discussed in UNDP, (2009), serves as a plan for project M&E and should clarify what is to be monitored and evaluated, the activities needed to monitor and evaluate, who is responsible for M&E activities, when M&E activities are planned (timing), how M&E activities are carried out (methods), what resources are required and where they are committed (UNDP, 2009).

UNDP (2009) asserts that relevant risks and assumptions in carrying out planned M&E activities should be seriously considered, anticipated and included in the M&E framework. Three common types of M&E frameworks are discussed in this study, namely the conceptual or narrative framework, the logical framework or the logic models and results framework (Frankel & Gage, 2007).

3.3 Logical framework

In the execution of projects, several factors contribute to the successful implementation of the project. Factors such as the resources allocated; financial, human and material being put to judicious use; and activities being followed through to ensure that resources reflect the output and outcome. Finally, the desired impact or goal is achieved. It is therefore vital to systematically plan the inputs and processes to achieve the intended project outcome. Crawford and Bryce (2003) opine that the aid industry leads in the use of the logical framework approach as a project design tool. However, its utilization has been limited to project financing and the operationalized M&E information systems to support project implementation. The widespread use of the logical framework, however, has not been without limitations. The major drawback of the logical framework approach was the assumption that certain external factors could not be controlled, therefore relying on assumptions.

The logical framework approach (LFA) was developed in the early 1960s by the USAID and NORAD in response to planning and monitoring of development projects (Barasa, 2014). A logical framework (LogFrame), also referred to as a logic model, provides a linear, logical interpretation of the relationship between specific input needs for carrying out planned activities to produce the specific outputs, resulting in the specific outcomes and impacts comparable to the objectives and goals of the project on the horizontal axis. The logical framework can be described as a planning and management tool (Barasa, 2014) for projects because it details the current progress of the project and where the project should be within a specific time frame. It further places much importance on the linkage between resources, activities, outputs and outcomes which serves as the basis to develop a comprehensive management plan (Gage & Dunn, 2009). The World Bank (2004) posits that the logical framework clarifies project objectives and assists in identifying the causal links and performance indicators at each stage of the resulting chain; thus, input – process – output – outcome – impact. The log frame obliges as a useful tool practically during implementation to take corrective actions and review progress (World Bank, 2004).

It is worth noting that a series of "if–then" relationships as shown in Figure 3.1 links the components of the logic model and underpins the vertical logic of the log frame: If resources are available to the programme, the programme activities can be implemented; if programme activities are implemented successfully, then

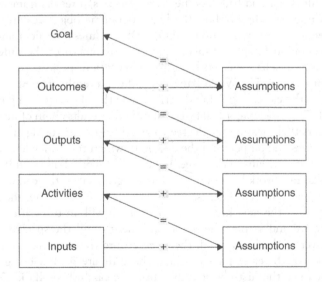

Figure 3.1 The "if–then" relationship that underpins the vertical logic of the log frame.

Source: Crawford et al., 2003

specific outputs and outcomes can be expected (Crawford, & Bryce, 2003; Gage & Dunn, 2009). Similar to a conceptual framework, the logic framework focuses on the project inputs, activities and results. This narrow focus assists programme managers and M&E planners as they clarify the direct relationships among elements of interest in a project (Gage & Dunn, 2009). It appears therefore that the logical framework concerns itself with assumptions that link project deliverables and cannot guarantee the desired project outcome owing to the complex and dynamic nature of the construction industry and the over-reliance on assumptions. Projects are unique to environment, stakeholders and beneficiaries and, as such, defy logic to achieve success.

3.4 Results framework

Results frameworks are also referred to in literature as strategic frameworks. A strategic framework establishes the direct relationship that exists or is expected to exist between the intermediate results of activities all the way to the overall objectives and goals. USAID (2013) indicated that strategic frameworks show the causal relationship between programme objectives and outline how each of the intermediate results/outputs and outcomes relates to and facilitates the achievement of each objective and how objectives relate to each other and the ultimate goal. Results frameworks form the basis for M&E activities at the objective level (Frankel & Gage, 2007). They ask questions of how we want the process to proceed at the initial stages of M&E, what the expectations are and, finally, why we want them.

It is obviously impossible to describe a project as successful if the project outcome or result is not laid bare. As the name suggests, a results framework (RF) presents an understanding of how the key programme objectives are achieved. The Independent Evaluation Group (IEG) (2012) defines a results framework as a clear presentation (graphic display, matrix or summary) of the different levels or chains of results expected from an intervention, project, programme or development strategy. The World Bank also defines a results framework as a representation of the causal logic that describes how the objective of a project is to be achieved. This is accomplished by translating the results chain of intervention into indicators that measure the degree to which inputs are being changed into specific activities and outputs and the degree to which the anticipated outcomes of the project are relevant to the target population (World Bank, 2013).

The results framework builds on the causal logic, that is, the cause and effect concept to achieve results. That is, if a lower objective is accomplished successfully, it will affect the next higher objective which will, in turn, affect the next higher objective and so on. The accomplishment of all the specific objectives from the lower to the highest will therefore ensure the achievement of project success. The RF therefore indicates how the ultimate goal is being achieved immediately once the first objective has been accomplished (IEG, 2012). This provides the M&E team with early warning signs of project progress relating to schedule, cost, quality and satisfaction and where interventions are required to

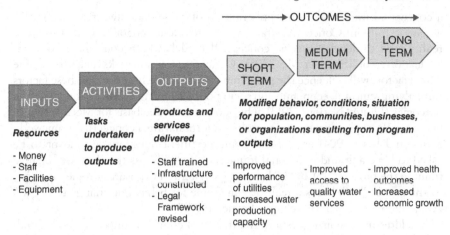

Figure 3.2 Results chain depicting how inputs are translated to achieve project outcomes.

Source: Adapted from the World Bank, 2013

address any challenge. Again, the results framework gives clarity to the theory of change, that is, it gives reasons why the programme will lead to the output and why the output will lead to an immediate and intermediate outcome, and further, why the outcome will lead to the long-term impact (goal) of the project with respect to a specific time frame (IEG, 2012).

The IEG (2012) postulates that the focus of the results framework is the outcome and impact of a project with minimal recognition to output. However, inputs and activities are not emphasized. This presents a conceptual results chain of output, outcome and impact, which is accompanied by a comprehensive plan for progress monitoring for impact through the measurement of output, outcome and impact at different intervals of the project life cycle (IEG, 2012). See Figure 3.2.

Results frameworks harness the resources available to achieve the desired results by engaging the inputs (the financial human and material resources used for development intervention) in the activities (actions taken through which inputs are mobilized to produce specific outputs) to generate outputs (the products, capital goods and services that result from development processes). The outputs generated therefore result in outcomes (the short-term and medium-term effects of a process's output) that produce a corresponding impact (actual or intended changes). Planning is an important feature of the results framework as everything will be subject to planning right from the onset of the process.

3.5 Conceptual or narrative framework

According to a release by PATH in 2011, conceptual frameworks are seen as diagrams that identify and illustrate relationships among relevant organizational, individual and other factors that may influence a programme and the achievement of goals and objectives. Stem, Margoluis, Salafsky and Brown (2005) defined a conceptual framework as a representation of cause-and-effect relationships in a generic fashion. Conceptual frameworks provide a generalized description of reality used to develop specific conceptual models which could be used as an evaluation tool (Stem et al., 2005). Conceptual frameworks help determine which factors will influence the programme and outline how each of these factors (underlying, cultural, economic and socio-political) might relate to and affect the outcomes (United Nations, 2013). They do not form the basis for M&E activities but can help explain programme results. In formulating a conceptual framework, Kusek and Rist (2004) argued that some pertinent questions are meant to be asked to have a fit and functional frame such as "What is the theory of change framing the process?" "What is the range of potential exposures people may have to the process?" "What is a realistic timeframe for behaviour change to occur?" And, finally, "how will this change be measured?"

In addressing the first question, the theory of change is important to be studied. As opined by Taplin and Clark (2012) and Taplin, Clark, Collins and Colby (2013), the theory of change reflects the underlying process and pathways through which the hoped-for change, which can be in the form of knowledge, behaviour, attitudes or practices, at the individual, institutional, community or another level, is expected to occur. A theory of change defines the pieces and steps necessary to bring about a given long-term goal (Stem et al., 2005). A theory of change describes whether it is a single programme or a comprehensive community initiative, i.e. the types of interventions that bring about the results hoped for. It includes the assumptions that stakeholders use to explain the process of change (Kusek & Rist, 2004). This is often supported by research (Stem et al., 2005).

The other questions raised lead to more questions such as "Who is going to be exposed directly to the process?" "Who will be receiving services?" and "Who might be exposed indirectly to the intervention?" The conceptual framework identifies appropriate measurements for the kind of change that is expected (Frankel & Gage, 2007; McDonald et al., 2007). The UNDP (2009) also indicated that a narrative framework should reflect the plans that may be in place to strengthen national or sub-national M&E capacities, the existing M&E capacities and an estimate of the human, financial and material resource requirements for its implementation.

3.6 Steps in developing a monitoring and evaluation framework

In recent times, researchers in the field of M&E have come up with numerous steps to be taken to formulate a functional and effective M&E framework. One of those works is seen in Frankel and Gage (2007). The steps are as follows:

1 Determine the purposes of the M&E mechanisms and assess the information needs of each actor;
2 Ensure that prevention and response interventions have defined objectives, outputs and indicators;

3 Establish coordinated and common reporting tools;
4 Determine methods for obtaining information on indicators;
5 Assign responsibilities for information gathering, determine the period and frequency of data collection and allocate resources; and
6 Establish mechanisms for sharing information and incorporating results into prevention and response planning.

3.7 A monitoring and evaluation system

Figure 3.3 illustrates an M&E system developed and used by the United Nations Development Program (UNDP, 2009: 55). According to Hardlife and Zhou (2013), programmes with the right technology and adequate financial supports are seen to be performing poorly owing to the lack of understanding of the balance between technology, management and capital. It is therefore imperative to institute an M&E system that will manage the entire programme process. An M&E system is a set of organizational structures, management processes, plans, indicators information systems, reporting lines and standards that ensure that projects are implemented effectively. The need to deliver projects to set objectives and standards has increased the consciousness among construction industry professionals about the relevance of an M&E system. The system aids in checking progress, ensuring efficient and effective utilization of project resources as well as the relationship between the project and M&E team. During M&E of projects, systems ensure that inputs, activities and processes are well-coordinated

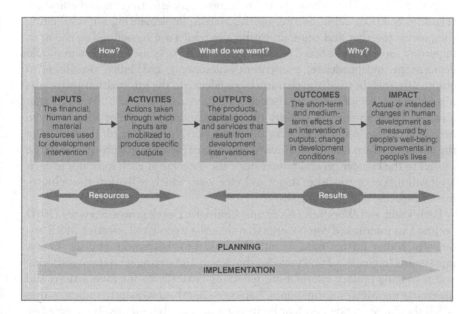

Figure 3.3 A monitoring and evaluation system.
Source: UNDP, 2009, p. 55

and checked regularly and consistently to warrant the desired output, outcomes and impact are met as planned. Furthermore, the system will also ensure the M&E activities are executed coherently as intended and will reveal all bottlenecks in the project implementation process (Kerzner, 2017). An M&E system also serves as a guide to facilitate the process of collection, analyzing and reporting on data based on agreed benchmarks of performance indicators (Omonyo, 2015). It serves the overarching accountability challenge in most project implementation. An M&E system is necessary to provide information that measures and guides the project plan, certifies processes, meets internal and external reporting requirements and informs future programming (Chaplowe, 2008). In developing a good system to monitor and evaluate projects and activities, it is imperative to consider who the primary beneficiaries are, the intended goal or desired change, the assumptions that link the project objectives to the particular intervention and the project scope and size (Chaplowe, 2008; Guijt & Woodhill, 2002; Ile et al., 2012).

The M&E systems not only serve as a guide to the implementation process but also inherently present the opportunity to gather relevant results information regarding progress for effective decision making (Briceño, 2010; Hardlife & Zhou, 2013; Njama, 2015) and also efficiently communicate such information to stakeholders to foster interaction among stakeholders to build team spirit (Routledge, 2015). Further, a clear responsibility for M&E is defined (Ile et al., 2012; Kusek & Risk, 2004). For the system to operate efficiently and generate the needed information, monitoring systems, as well as evaluation systems must complement each other to maximize their strengths and weaknesses for effectiveness and efficiency (Hardlife & Zhou, 2013). Setting up the system, implementing and evaluating the process together and communicating the results are recognized as the most important aspects of the M&E process (Njama, 2015). As such, the system should demonstrate independence, transparency, autonomy, credibility, usefulness to ensure sustainability and also guarantee confidence in the information and results it generates for decision making (Briceño, 2010). Also, Briceño (2010) asserts that a proper M&E system will identify the impact and challenges of the project to help project managers strategize for future projects. In this regard, the success of the project is dependent on the interrelatedness of every stage (sub-system) of the project to the broader general system since the outcome at each indicator at each level will influence the other stages and vice versa which will affect the achievement of specific and general goals.

Both Guijt and Woodhill (2002) and Umhlaba Development Services (2017) outline four interlinked components that describe a good and effective M&E system which they further indicate must be linked to the project strategy. This is represented in Figure 3.2 and includes the design and setting up system, gathering and managing project information, reflecting critically to improve actions and disseminating M&E information. Chaplowe (2008) subsequently indicates that M&E should be founded on four main components: that change is expected by the project, the precise objectives leading to the change, indicator and how they will be measured and, finally, how data will be collected and analyzed. In sum, the system aids management to take decisions on the overall success, failure, relevance, efficiency and effectiveness of their programmes (Hardlife & Zhou, 2013).

3.8 Types of a monitoring and evaluation system

Several types of M&E systems exist and have been developed by donor agencies for use by programme and donor partners to ensure programmes and interventions are executed as planned towards achieving the programme objectives. Developing M&E systems will depend on several factors including the focus, need and expectation of the programme or project being implemented. It is averred that there is lack of a universal M&E model; hence systems are tailored to suit the type, complexity and size of the programme, institutional setup, managerial responsibility, the reporting requirement and the frame conditions of the project or programme (Gudda, 2011). Kusek and Rist (2004) identify two main types of M&E systems which are presented in the logic flow (Figure 3.1). These categories are recognized as implementation-based M&E systems and the results-based M&E systems and are explained below.

3.8.1 Implementation-focused M&E system

Traditionally, M&E are conducted to address compliance issues; thus to address the 'Did they do it?' question (Kusek & Rist, 2004) and 'Was the project delivered on time?', 'Did the project exceed the planned budget?' and 'Was the project of the right and approved quality? From Figure 3.4, the implementation of M&E is focused on how well the project or programme is being executed. Unfortunately, this approach to M&E provides little information to stakeholders and the M&E team on the understanding of how the project achieved success or failure. The data collected during the implementation-focused M&E covers the inputs that have been provided, activities being undertaken and the output as seen. Also, M&E reports capturing the provision and utilization of the project inputs and the production output.

3.8.2 Results-based M&E system

The results-based M&E system is a work-in-progress system that is used by project and programme implementers to track progress and validate the impact of the project or programme. Greater focus is on the achieved outcome and impact of the project to the beneficiary community or end users. To develop and maintain an effective results-based M&E system, the need for continuous commitment of time, efforts and resources is important (Kusek & Rist, 2004). Also, political, organizational and technical barriers need to be overcome to ensure the system works effectively to deliver successful projects. According to Kusek and Rist (2004), some international initiatives have been introduced to force governments and donor partners to adopt systems which are geared towards result achievement. An example is the sustainable development goals (SDGs). Additionally,

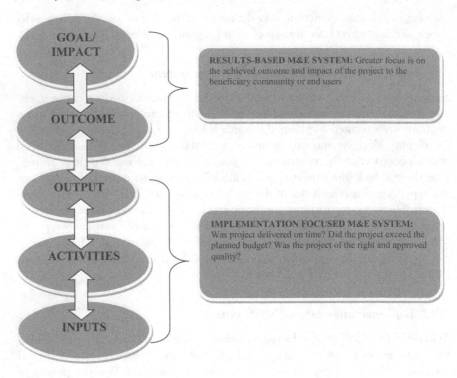

Figure 3.4 Flow chart of the M&E system.

Source: Researcher

for instance, the schools under trees and emergency intervention programme by the government of Ghana require being monitored and evaluated to ascertain whether the initiative is making any difference and having an impact on the end users compared to the initial situation before the project was undertaken. It is therefore important to develop a results-based M&E at the organizational level (MMDAs) to measure and monitor the achievement of infrastructural development initiated by the government, donor organizations, NGOs and even internally generated funded projects.

3.9 Criteria for assessing the quality of a monitoring and evaluation system

The need to have a standard measure of an M&E system for construction project delivery is significant (Tengan, Aigbavboa & Oke, 2018). The literature on the criteria for determining a suitable M&E system has, however, not established one single accepted standard for evaluating the quality of M&E systems. However, the quality and success of an M&E system are linked to the quality of the M&E indicators that have been formulated (Sharma, 2010). The International Fund for

Agricultural Development (IFAD) (2002) outlines four broad criteria for assessing the quality of a project monitoring and evaluation system, namely utility, feasibility, propriety and accuracy. The M&E system must be useful and helpful to serve the practical information needs for which it is developed. Feasibility concerns the methods, sequences, timing and processing procedures proposed being realistic, prudent and cost-effective, while in terms of propriety, the M&E activities will be conducted legally, ethically and with due regard to the wellbeing of those affected by its results. Finally, the system will generate data to be analyzed to inform the decision for improvement. The exactness of such data is necessary for reliable decision making (Chaplowe, 2008).

3.10 Steps in developing a monitoring and evaluation system

Several studies show varying numbers, steps and specific sequences for developing an M&E system (Kusek & Rist, 2004). Whereas different scholarly works have recommended between four and seven steps (Motingoe, 2012), Kusek and Rist (2004) outline ten steps for designing an M&E system. This indicates that there is no single approach to designing and developing an M&E system. However, according to Motingoe (2012), owing to project circumstances and the unique features of projects, M&E systems must be developed on a case-by-case basis. While professing the ten-step approach to designing an M&E system, Görgens and Kusek (2009) further indicate that significant strategies and activities must be documented and grouped logically as well as finalized in an appropriate sequence. The proposed ten steps by Kusek and Rist (2004: 25) in designing an M&E system is summarized below and illustrated in Figure 3.5.

1 Conducting a readiness assessment;
2 Agreeing on outcomes to monitor and evaluate;
3 Selecting key indicators to monitor outcomes;

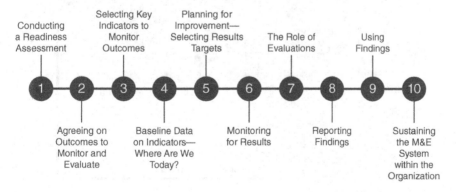

Figure 3.5 Steps in designing, building and sustaining a results-based M&E system.

Source: Kusek and Rist, 2004

4   Determining a baseline survey of indicators – where are we today?
5   Planning for improvement – selecting results targets;
6   Monitoring for results;
7   Defining the role of evaluation;
8   Reporting findings;
9   Using findings; and
10 Sustaining the M&E system within the organization.

Figure 3.6 presents an illustrative summary of how an M&E system is designed. Even though it has been presented in a linear or sequential format, it does not represent the same during its implementation (Motingoe, 2012). According to Kusek and Rist (2004), it may be necessary to work back and forth and not necessarily in a strict sequence to ensure systems are robust and effective. A combination or expansion of some of the steps outlined in the Kusek and Rist model may be necessary for individual projects, depending on the purpose and utility of the M&E system. M&E systems, however, have been criticized for generating information that is delivered and received late and for not answering the right questions. They are also seen to be expensive to implement (Motingoe, 2012). Guijt and Woodhill (2002) also outlined four steps in an M&E process and detailed how it is linked to the project strategy and operations. These steps are illustrated in Figure 3.6 and involve developing the M&E system, gathering and managing information, reflecting critically to improve actions and communicating and reporting results.

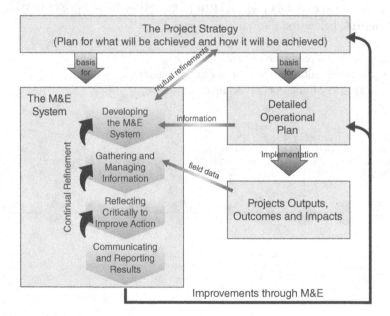

Figure 3.6 The M&E process and how it links to the project strategy and operations.

Source: Guijt and Woodhill, 2002

3.11 Components of a monitoring and evaluation system

M&E systems are broken down into interrelated and interconnected components to allow special attention to be provided by the project M&E team. This will ensure that the M&E system can collect and analyze the data, report the findings and take decisions for improvements. Several components of an M&E system are identified in the literature. While Ile et al. (2012) demonstrate two complementary parts of an M&E system, a six-component system has been professed by the FAO and includes a clear statement of measurable objectives for the project and its components; a structured set of indicators covering project inputs, process, outputs, outcomes, impact and exogenous factors; mechanisms for data collection with a capacity to monitor progress over time and baselines; a method to compare progress and achievements against targets; appropriate building on baselines and data collection with an evaluation framework and methodology capable of establishing causation; clear mechanisms for reporting and using M&E results in decision-making; and, finally, sustainable organizational arrangements for data collection, management, analysis and reporting.

Similarly, the Independent Evaluation Group (IEG, 2013) of the United Nations Population Fund (UNPF) describes four components of an M&E system. They are outlined as the monitoring of inputs and activities, the monitoring of outputs and outcomes, monitoring of risk assumptions and, finally, evaluation of the entire process (IEG, 2013). Görgens and Kusek (2009) and the UNAIDS (2009) have both outlined twelve components of an M&E system. It is suggested by Mtshali (2015) that the twelve components are a conglomeration of many other components discussed by other scholars. The twelve components of the M&E system are further categorized into three: the first six components are categorized as "people, partnering and planning" (outermost components), the second category comprises components seven to eleven which are attributed to "collecting, capturing and verifying data" (middle component) and the third category relates to "using data for decision-making" which refers to the twelfth component of the M&E system found at the core/centre of Figure 3.5 (Mtshali, 2015). The three categories of M&E components are discussed and illustrated in Figure 3.7.

3.11.1 Category one

Category one comprises six components of M&E. They are organizational structures with M&E functions; the human capacity for M&E; a partnership for planning, coordination and managing the M&E system; the M&E frameworks/logical framework; the M&E work plan and costs and communication; and advocacy and culture for M&E. The need for an established M&E unit with the aim of coordinating all monitoring and evaluation functions at all levels of project implementation is advocated. While M&E services can be outsourced to external project management consultants, there is still the need for an internal unit within the organization to oversee its M&E functions. It is imperative therefore to have within the organization adequate staff to oversee M&E functions if the effective

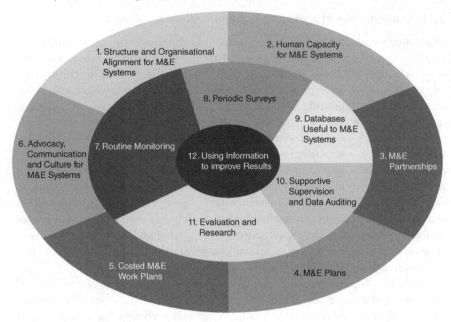

Figure 3.7 Components of the M&E system.

Source: Görgens et al., 2009; UNAIDS, 2009

M&E implementation is expected. Staff's M&E technical know-how, knowledge and experience are critical and this can be enhanced through continuous M&E capacity development training. This component also describes the level of partnership as a pre-requisite for successful M&E systems. Partnerships complement the M&E efforts of the organization. This is seen in the partnerships that exist among the local, regional and national planning commissions. They also serve auditing purposes where line ministries, technical working groups, communities and other stakeholders can compare M&E outputs with reported outputs.

An M&E framework is needed to describe the objectives, inputs, outputs and outcomes of the proposed project and the pointers that will be used to measure their achievements. Without the framework, project performance cannot be measured to ensure projects are on track or otherwise. Similarly, the M&E work plan and associated cost are important to outline how the resources allocated for the M&E functions will be used to achieve the goals of M&E. The work plan shows how personnel, time, materials and money will be used to achieve the set M&E functions. Finally, the communication of M&E results, advocacy and the culture for M&E refer to the incidence of policies and strategies within the organization to promote M&E functions. Communication and strategies need to be supported by the organization's hierarchy. The existence of an organizational M&E policy, together with the continuous use of the M&E system outputs on communication channels, are some of the ways of improving communication, advocacy and culture for M&E.

3.11.2 Category two

Five components comprise the second category of the M&E. Routine M&E refers to the continuous and routine data gathering during project implementation. To indicate whether project activities are producing the right indication of meeting project objectives requires the continuous collection of project data or information. Surveys and surveillance from a higher M&E body such as the regional or national planning commission need to be continuous to evaluate the progress of related projects. For example, construction project delivery initiated by the Ghana Education Trust Fund (GETFund) at the local government level will require survey and surveillance by the GETFund to help measure the level of success of the implemented project.

The need for the national and sub-national database is imperative as the world is becoming an open source. Partnership with funding organization would require relevant data. Hence, M&E systems need to develop strategies for submitting appropriate, dependable and valid data to national and sub-national databases. Incorporating data auditing infers that the data is subjected to verification to certify its reliability and validity. Supervision corroborates the auditing process and is important as it ensures the M&E process is run efficiently. Evaluation and research are important components of the M&E process as they establish whether the desired objectives of projects have been met. They usually provide for organizational learning and sharing of successes with other stakeholders.

3.11.3 Category three

Finally, disseminating and utilizing M&E data generated from project implementation for the benefit of future project management is critical. It helps either to reinforce, revise or change the implemented strategy. Moreover, both monitoring and evaluation results need to be made available to key stakeholders for accountability purposes.

3.12 A monitoring and evaluation plan

M&E plan development is an indispensable tool to manage the process of assessing and reporting progress towards achieving project success and to identify what evaluation questions need to be addressed through evaluation. Frankel et al. (2007) describe an M&E plan as a central document which explains in detail the programme's objectives and the interventions developed to achieve these goals. It also describes the procedures that will be implemented to ensure the goals are realized. An M&E plan is necessary for every project. However, the stage or period to develop the plan is in contention. Whereas Omonyo (2015) asserts that an M&E plan should be developed at the planning stage as part of the project initiation document (PID), Frankel et al. (2007) are of the view that the M&E plan should be created after the planning stage but before the design phase of the project. Key components of the M&E plan, according to Badom (2016), consist of a clear, measurable and achievable objective based on which indicators will be defined. Secondly, the M&E plan contains a SMART set of indicators covering all the elements of the project, namely inputs, process or activity, output, outcome and impact. The M&E plan also makes a provision for the collection and management of project information, an arrangement for gathering and analyzing project data and a proposed mechanism for generating feedback for decision making (ibid).

Summary

The chapter discussed M&E systems and frameworks critical for effective planning and management of the project cycle, providing a clear understanding of the M&E process and the relationship that exists between project activities. A detailed description and illustration of key M&E frameworks and systems are discussed in reference to construction project delivery. Criteria for assessing the quality of an M&E system, steps in developing an M&E system as well as the components and plans of an M&E system were at the center of this chapter. The next chapter under part II of this book will consider the theories, models and concepts in M&E research.

References

Al-Otaibi, M. (2011). *Evaluation of contractor performance for pre-selection in the Kingdom of Saudi Arabia.* PhD Thesis. UK: Loughborough University.

Barasa, R. M. (2014). *Influence of monitoring and evaluation tools on project completion in Kenya: A case of Constituency Development Fund projects in Kakamega County, Kenya.* University of Nairobi, Kenya.

Briceño, G. (2010). *Defining the type of M&E system: Client, intended users, and actual utilisation.* Washington, DC: World Bank Publications.

Chaplowe, S. G. (2008). *Monitoring and evaluation planning: Guidelines and tools.* Washington, DC and Baltimore, MD: Catholic Relief Services.

Crawford, P. & Bryce, P. (2003). Project monitoring and evaluation: A method for enhancing the efficiency and effectiveness of aid project implementation. *International Journal of Project Management,* 21(5), pp. 363–373, doi:10.1016/S0263-7863(02)00060-1

Frankel, N. & Gage, A. (2007). *M&E fundamentals: A self-guided minicourse.* MEASURE evaluation, Interagency Gender Working Group, Washington D.C.: U.S Agency for International Development.

Gage, A. & Dunn, M. (2009). *Monitoring and evaluating gender-based violence prevention and mitigation programs.* MEASURE Evaluation, Inter-agency Gender Working Group. Washington DC: US Agency for International Development.

Görgens, M. & Kusek, J. Z. (2009). *Making monitoring and evaluation systems work – A capacity development toolkit.* Washington: World Bank.

Gudda, P. (2011). *A guide to project monitoring & evaluation.* United States of America: AuthorHouse.

Guijt, I. & Woodhill, J. (2002). *Managing for impact in rural development: A guide for project M&E.* IFAD: Office of Evaluation and Studies (OE).

Hardlife, Z. & Zhou, G. (2013). Utilisation of monitoring and evaluation systems by development agencies: The case of the UNDP in Zimbabwe. *American International Journal of Contemporary Research,* 3(3), pp. 70–83.

Ile, I. U., Eresia-Eke, C. & Allen-Ile, C. (2012). *Monitoring and evaluation of policies, programmes and projects.* Hatfield, Pretoria: Van Schaik.

Independent Evaluation Group (IEG). (2012). *Designing a result framework for achieving results: A how-to guide.* Washington, DC: World Bank.

Independent Evaluation Group (IEG). (2013). *Assessing the country office monitoring and evaluation system.* Washington, DC: World Bank.

Kerzner, H. (2017). *Project management: A systems approach to planning, scheduling, and controlling.* John Wiley & Sons Inc., Hoboken, New Jersey

Kusek, J. Z. & Rist, R. C. (2004). *Ten steps to a results-based monitoring and evaluation system: A handbook for development practitioners.* Washington, DC: World Bank.

McDonald, K. M., Sundaram, V., Bravata, D. M., Lewis, R., Lin, N., Kraft, S. A., McKinnon, M., Paguntalan, H. & Owens, D. K. (2007). *Conceptual frameworks and their application to evaluating care coordination interventions.* Rockville, MD: Agency for Healthcare Research and Quality (US). No. 9.

Motingoe, R. S. (2012). *Monitoring and evaluation system utilization for municipal support.* PhD Thesis, NorthWest University, South Africa.

Mtshali, Z. (2015). *A review of the monitoring and evaluation systems to monitor the implementation of early childhood development within Gauteng Department of Health.* Master's dissertation. Stellenbosch: Stellenbosch University.

Njama, A. W. (2015). *Determinants of effectiveness of a monitoring and evaluation system for projects: A case of Amref Kenya WASH programme.* Nairobi: University of Nairobi.

Omonyo, A. B. (2015). *Lectures in project monitoring & evaluation for professional practitioners.* Germany: Lambert Academic Publishing.

Routledge (2015). *Routledge handbook of higher education for sustainable development.* New York: Routledge.

Sharma, R. (2010). Challenges in monitoring and evaluation: An opportunity to consolidate the M&E systems. In: *Challenges in Monitoring and Evaluation: An Opportunity to Institutionalize M&E Systems. Presented at the Fifth Conference of the Latin America and the Caribbean Monitoring and Evaluation (M&E) Network.* Washington, D.C.: The World Bank.

Stem, C., Margoluis, R., Salafsky, N. & Brown, M. (2005). Monitoring and evaluation in conservation: A review of trends and approaches. *Conservation Biology*, 19(2), pp. 295–309.

Taplin, D. H. & Clark, H. (2012). *Theory of change basics. A primer on theory of change.* New York: ActKnowledge.

Taplin, D. H., Clark, H., Collins, E. & Colby, D. C. (2013). *Theory of change.* New York: Center for Human Environments.

Tengan, C., Aigbavboa, C. O. & Oke, A. E. (2018). Evaluation of UFPA quality assessment criteria for monitoring and evaluation system in the Ghanaian construction industry. *African Journal of Science, Technology, Innovation and Development*, pp. 1–5, doi:10.1080/20421338.2017.1423008

Umhlaba Development Services (2017). *Introduction to monitoring and evaluation using the logical framework approach.* Johannesburg, South Africa: Umhlaba Development Services.

UNAIDS (2009). 12 Components monitoring and evaluation system assessment. Joint United Nations Programme on HIV/AIDS, Geneva

United Nations (UN) (2013). *Ending Violence against Women and Girls Programing Essentials.* United Nations Entity for Gender Equality and the Empowerment of Women.

United Nations Development Programme (UNDP) (2009). *Handbook on planning, monitoring and evaluation for development results*. New York, USA: UNDP.

United States Agency International Development (USAID) (2013). Developing Results Framework. Version 1. Technical Report. Bureau for Policy, Planning and Learning

World Bank (2004). *Monitoring & evaluation: Some tools, methods and approaches*. Washington, D.C.: Operations Evaluation Department (OED).

World Bank (2013). *Results framework and M&E guidance note*. Washington, DC: World Bank

Part II

Theories, models and concepts in monitoring and evaluation research

4 Theories of monitoring and evaluation

4.1 Abstract

Evaluation theories serve as the fulcrum or knowledge base of the evaluation practice helping to shape the present and future practices of evaluation. However, due to the difficulty of earlier evaluation theories to respond to the initial mandate of evaluation needs, several evaluation theories were advanced. The need for guidance on how evaluation should be done was prominent in the early years of the development of the evaluation field which saw prominent evaluators prescribing what they believed was the way to conduct an effective evaluation. The evolution of the evaluation theory tree (ETT) prescribes social accountability and enquiry and epistemology as the foundation for the development of evaluation. Further, three main paradigms that guide the evaluation practice are method, value and use. The method branch of evaluation is founded on the positivist social science paradigm which focuses on realising the objective truth about the causes and effects of programmes and the generalisation of findings, whereas the value branch of evaluation is theoretically founded on the constructivist paradigm which suggests that reality is socially constructed and that knowledge is also created by our own experiences. Finally, the "use" branch focuses on how and who will use the evaluation findings. The understanding of these theories will guide the development of monitoring and evaluation systems to achieve specific needs of evaluation.

4.2 Introduction

Monitoring and evaluation has specific objectives to achieve. The focus of M&E is to guarantee project success by ensuring that projects meet cost, schedule, quality and satisfaction targets (Tengan & Aigbavboa, 2016). However, achieving project success is tedious and complicated (Berssaneti & Carvalho, 2015; Tengan & Aigbavboa, 2016). Implementing M&E ensures resources and processes/activities are efficiently and effectively utilized to ensure the right output and outcomes are achieved as well as the achievement of project impact. The likelihood of continually achieving project success will largely depend on how systematically and continuously the monitoring process is undertaken and the interplay of the relationship of various independent factors. Again, achieving M&E objectives requires evaluators to understand why the need for the evaluation, who will make use of the evaluation findings and how the evaluation would be done. It is only after such understanding is gained that true meaning will be given to the practice.

4.3 Monitoring and evaluation theory defined

The origin of M&E is rooted in the perception of public sector failures as early as the 1950s (Cameron, 1993). However, the concept of evaluation occurred in the USA in the 1960s and 1970s with support from the federal government under the strategies of "war on poverty" and "the Great Society" (Waithera & Wanyoike, 2015). Some projects underperformed (failed) which led to the creation of an independent M&E unit with responsibilities ranging from collecting data, processing and analyzing the same (Cameron, 1993). The M&E departments' responsibilities also included reporting on project performance against original targets to ministries and international agencies. M&E has been around for a while and has featured in many disciplines of study and practice. These include finance, governance, agriculture, development project and health. M&E for accountability and systematic social enquiry has been the focus or truck of evaluation research (Alkin & Christie, 2004). Notwithstanding the long-standing profession and availability of literature on the subject matter, the ultimate object and impact in project delivery can be questioned since projects do not conform to quality standards, cost overruns and completed beyond schedule and general dissatisfaction by project stakeholders.

M&E as a project management function is a key driving factor for achieving project success. As noted by Kamau and Mohamed (2015), M&E is a critical success factor (CSF) in project delivery. Achieving project success (PS) is nearly impossible without the constant M&E which has been found consistent with literature and practice (Ika et al., 2012; Kamau & Mohamed, 2015; Papke-Shields et al., 2010; Prabhakar, 2008). Also Kibebe and Mwirigi, (2014) ranked ineffective M&E as the number one management factor that contributed to project failure. The above reiterate the importance of M&E in the construction industry. Even though M&E are two management functions undertaken closely together to achieve stated goals and objectives of projects, two schools of thought exist regarding their relationship. Whereas M&E are seen as two separate functions by many evaluation scholars, others have argued the inseparable nature of the two management functions (Musomba, Kerongo, Mutua & Kilika, 2013; Omonyo, 2015).

4.4 A review of monitoring and evaluation theories

A theoretical approach to M&E can be described as a set of knowledge which helps understand the study and practice of M&E from several viewpoints (Waithera & Wanyoike, 2015). The proliferation or advancement of several evaluation theories can be attributed to the difficulty of earlier evaluation theories to respond to the initial mandates of evaluation needs (Smith, 1993) as well as the need to guide the evaluation practice (Carden & Alkin, 2012). The theory defines a body of knowledge that organizes, classifies, describes, predicts or helps in understanding and controlling a topic (Shadish, Cook & Leviton, 1991). Theory can be prescriptive, thus a hypothesis of what ought to be done or a body of knowledge (Omonyo, 2015).

4.4.1 The evaluation theory

According to Shadish (1998), evaluation theory is the knowledge base of the evaluation profession and needs to be guarded seriously by all evaluation professionals. He further posits that evaluation theory is not concise or axiomatic and not a single theory but rather a collection of diverse theoretical writings held together by the common glue of having evaluation practice as their target. Evaluation theories can be very informative for initial needs assessment and programme design. Evaluation theory gives effective strategies for dealing with the challenges regarding the evaluation process. Lessons are learned about what doesn't work which may save programme designers' and evaluators' time and resources (Donaldson, 2001). According to McCoy et al., (2005), evaluation theory compares the project impact with what was set to be achieved in the project plan, thereby assessing effectiveness in achieving project goals and in determining the relevance and sustainability of an ongoing project. Shapiro (2004) informs that two forms of evaluations exist, depending on when it is undertaken. These are formative and summative evaluations. Formative evaluation (interim or mid-term evaluation) is concerned more with efficient use of resources to produce outputs and focuses on strengths, weaknesses and challenges of the project and whether the continuous implementation of the project plan will be able to deliver the project objectives or whether it needs redesigning (Passia, 2006). A summative evaluation is undertaken at the end of the project and aims at determining how the project progressed and what went right and wrong and capturing any lessons learned.

Evaluation theory is geared towards guiding future projects by facilitating organizational learning through the documentation of good practices and errors. There are some critical factors that need consideration during evaluation. These include the use of relevant skills, sound methodology, adequate skilled personnel, financial resources and transparency to ensure the effectiveness and quality of M&E (Jones et al., 2009). Rogers (2008) suggests the use of multi-stakeholders' dialogues in data collection, hypothesis testing and, in the intervention, to allow greater participation and recognize the differences that may arise. However, one of the limitations of evaluation theory is that for any evaluation process for projects to be successful, it must be done within a supportive institutional framework while being cognizant of "political" influence.

4.4.1.1 The evaluation theory tree

The evaluation theory tree, as opined by Alkin and Christie (2004), described in detail (see Figure 4.1) the inputs of various early contributors, theorists and researchers to the theory of evaluation. The fulcrum around which evaluation has gained recognition are the social accountability, social inquiry and epistemology of evaluation which have been referred to as the root of evaluation (Carden & Alkin, 2012). Alkin and Christie (2004) and Carden and Alkin (2012) posit that the need for guidance on how evaluation should be done was prominent in the early years of the development of the evaluation field which saw prominent evaluators prescribing what they believed was the way to conduct the effective evaluation. The prescriptions on how the evaluation was to be conducted were referred to as **theories** while the prescriber assumed the title **theorist**. This, however, meant that such works had to have been fully developed and have stood the test of time before they could be recognized as a theory (Carden & Alkin, 2012; Omonyo, 2015).

The modified Carden and Alkin's evaluation theory tree professes three main paradigms; they maintained that evaluators should consider the **"user"** of the evaluation effort, the **methodology** being used and the way in which the evaluation data will be **valued** (Alkin & Christie, 2004; Carden & Alkin, 2012). However, the three branches are not independent of each other but rather they complement each other for a common purpose. The relationship between the branches presents an opportunity to utilise the strengths and opportunities of each approach as a complement for the best M&E practice (see Figure 4.1). The placement of theorists on the branches of the evaluation tree is influenced by the emphasis made by theorists in their works and their philosophies toward evaluation. A proposed additional evaluation perspective known as **"context"** defined the characteristics of the situation of the low- and middle-income countries

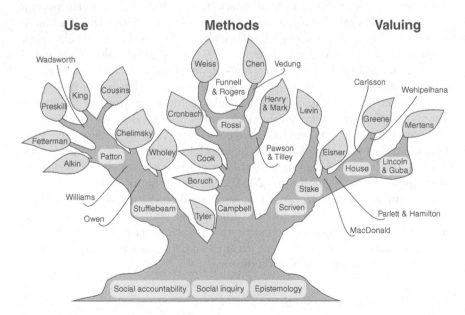

Figure 4.1 Modified evaluation theory tree.

Source: Adopted from Carden et al., 2012

Table 4.1 Summarized evaluation approaches in LMIC

	Use	Method	Value
Adopted	Logical Framework Analysis (LFA)/Results-Based Management (RBM). *Many bilateral development agencies as well as multilateral development banks (e.g. World Bank)*	3IE: International Initiative on Impact Evaluation Centre for Global Development	
Adapted	Developmental Evaluation *Patton*		Rapid Rural Appraisal, Participatory Rural Appraisal *Chambers*
	Outcome Mapping *Earl, Carden & Smutylo*		
	Most Significant Change *Davies and Dart*		
Indigenous	Citizen Report Card *Public Affairs Centre, India*		Systematization *Selener, Tapella et al.*
	African Peer Review Mechanism *New Partnership for Africa's Development*		

Source: Adapted from Carden & Alkin, 2012

(LMIC) (Carden & Alkin, 2012). The study, *Evaluation Roots: An International Perspective*, alluded to reasons that suggested the existing perspectives were the brainchild of North America, Australia, New Zealand and Europe and, as such, the evaluation field had developed based on the needs of these regions at the expense of other LMICs. The introduction of the "context" viewpoint was to ensure an international perspective of the theory tree (Carden & Alkin, 2012).

The study by Carden and Alkin (2012) revealed that several evaluation approaches are indigenous, adopted or adapted from North America and Europe into low- and middle-income countries (LMIC). The methodologies are placed on respective branches on the evaluation tree presented in Figure 4.1 as a modified evaluation tree. These are summarized and presented in Table 4.1.

4.4.1.2 *Method perspective*

The method branch is the central branch of the evaluation tree. This branch of evaluation is founded on the positivist social science paradigm which focuses on realising the objective truth about the causes and effects of programmes and concentrating on the generalisation of findings (Alkin & Christie, 2004). This evaluation approach recognises the existence of research and therefore is guided by research methodology (Carden & Alkin, 2012). Shadish et al. (1991) and Katz, Newton, Shona, & Raven (2016) further advance that the theorist on the method branch of the evaluation tree is focused on rigorous knowledge construction within the evaluation constraints and holds that reality exists independent of the observer and that the distance from the object of study helps avoid bias. Notable among methodological theorists is Cook whose work in the year 2000 indicated that evaluators who used qualitative methods could not establish causality between the observed outcome and the project itself since causality can only be established through experimental design (Omonyo, 2015). It is acknowledged among theorists that Tyler's work in the 1940s on the theoretical views as well as his attention on educational measurement was very prominent. Other renowned theorists whose work advanced methodological perspective in the evaluation included Peter Rossi, Thomas Cock, Donald Campbell and Edward Suchman. The objective of the method perspective of the evaluation theory tree emphasises effective M&E data collection and the utilisation of data to generate knowledge for management use.

4.4.1.3 Value perspective

The value branch of evaluation is theoretically founded on the constructivist paradigm which suggests that reality is socially constructed and that knowledge is also created by our own experiences (Mertens & Wilson, 2012). The value approach is positioned to the right side of the method branch and advances that the essence of evaluation is to facilitate value judgement and, as such, much emphasis is placed on the value of their findings in making judgements through the appropriate evaluation of outcomes (Carden & Alkin, 2012; Christie & Alkin, 2013). Micheal Scriven is noted to have led this perspective with his work in 1967 which largely proclaimed that evaluation without valuing is not evaluation (Carden & Alkin, 2012). Therefore, it is essential for evaluation to assess value objectively (Stufflebeam & Coryn, 2014). Theorists such as Robert Stake greatly influenced the value perspective. Omonyo (2015) asserts that the writing by Stakes in 1975 inspired evaluators to attend to actual programme activities rather than intents and presented different value perspectives when reporting on the success and failure of a project. The value paradigm argues the need to focus on project activities during M&E. Ensuring that activities scheduled receive the needed attention calls for the right representation of the stakeholders during the M&E who are equipped with the right technical capacity. The value branch was subsequently split into the objective and subjective sub-branches.

4.4.1.4 Use perspective

Finally, the use branch is located at the left of the method branch of the evaluation theory tree (see Figure 4.1). Alkin and Christie (2004) postulate that the use branch focused on how and who will use the evaluation findings. Theorist Stufflebeam is well-known as leading the crusade regarding the "use" paradigm. His work was primarily oriented toward management, decision and accountability and was based on the context, input, process and product (CIPP) model. This group of theorists was referred to as decision-driven and they placed much emphasis on conducting an evaluation to assist key programme stakeholders in programme decision-making (Katz et al., 2016). Patton's utilisation-focused evaluation model in 2008 is a classic example of works which focused less on the various methods used in the evaluation and rather more on the utility of the evaluation, stressing that evaluations should be designed to achieve findings that can be used by policy-makers to develop and improve policies and programmes (Katz et al., 2016). Originally, theorists on this branch focused on those individuals who were contracted to undertake evaluation and those empowered to use the information but this has been expanded to include a broader user audience and to evaluation capacity-building within the organization being evaluated (Alkin & Christie, 2004; Carden & Alkin, 2012). Patton's need for effective utilization of evaluation findings in decision- and policy-making resulted in the development of standards to ensure that evaluation findings are reliable and influence decisions for improvement. These he outlines as utility (ensure relevance and use); feasibility (realistic, prudent, diplomatic and frugal); propriety (ethical, legal and respectful) and accuracy (technically adequate to determine merit or worth) (Patton, 2008).

4.5 Theory of change

The theory of change (ToC) approach first emerged in the USA as early as the mid-1990s with the aim of improving evaluation theory and practice (Stein & Valters, 2012). Theory of change did not only seek to improve evaluation theory and practice but also evolved because of the emphasis placed on outcome and accountability by project and programme funders (James, 2011; Stein & Valters, 2012). This reason placed much attention on the need for a theory of change. As postulated by Weiss (2004), the ToC conceived that a key reason why a complex project fails is that the underlying expectations of the project are poorly articulated (Omonyo, 2015). Varied definitions exist regarding what theory of change stands for in different organizations in the evaluation field (Stein & Valters, 2012). A theory of change is described as a critical thinking process or tool that provides a comprehensive outlook of the early and intermediate changes that are needed to achieve a long-term goal or objective. Taplin and Clark (2012) assert that theory of change is a rigorous and participatory process in planning for long-term goals of a project by stakeholders through the identification of conditions or interventions which are believed will cause the achievement of the project long-term objective. In other descriptions, Rogers (2014) advanced that a theory of change is a set of philosophies describing the expected change, how the process will occur, what makes it happen and what must be done for the intended results to be achieved. This he further acknowledges as the enviable task of project stakeholders. The theory of change model is presented in a graphic form (see Figure 4.2) and outlines the immediate conditions as intended outcome in a casual framework or pathway (Stein & Valters, 2012; Taplin & Clark, 2012). Theory of change can therefore be explained as assumptions about how direct interventions contribute to achieving project success (Weiss, 1995).

It is believed that the theory of change approach has been adopted in many fields predominantly by the international non-governmental organization for

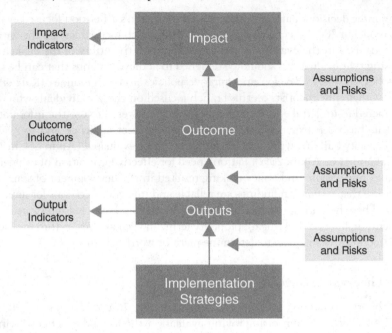

Figure 4.2 Schematic depiction of a theory of change.

Source: Adapted from Rogers, 2014

evaluation purposes (James, 2011). Stein and Valters (2012) indicate however that the use of the theory of change in the field of community development has become the mainstream owing to the demands of key funders as well as the desire to achieve some level of results by organizations. The purpose of the theory of change to organizations and project funders varies (Taplin, Clark, Collins, & Colby, 2013). Stein and Valters (2012) outline four purposes of a theory of change: Firstly, it is a support in systematic planning which integrates with the log frame to develop the process (pathways) leading to change in the intended outcome for project implementation; secondly, monitoring and evaluation aids in reviewing intended processes and outcomes over a period of time; thirdly, it provides a description which permits organizations to communicate selected changes of processes to partners; and the final purpose is learning which concerns the application of the theory of change as a thinking tool. Taplin et al. (2013) also identify that a theory of change drives communication through outcome pathways and narrative to stakeholders, builds core capacities, clarifies the relationship among stakeholders, strategizes influence on the boundary of stakeholders, plans outcome-based activities (approach to M&E) and clarifies M&E priorities.

4.6 Program theory

Program theory of evaluation has grown in use over the past decade. It assesses whether a programme is designed in such a way that it can achieve its intended outcomes. The programme theory is a guidance theory in the evaluation of projects as it shows the capacity of the programme to attend to specific problems within projects that need to be reviewed. It further offers guidance on what areas need to be emphasized during the evaluation process (Donaldson, 2001). In this book, programme theory is used because it presents the advantage of offering information that could lead to additional explanations regarding the M&E tools, employee training, management influence on M&E systems and stakeholder participation in project delivery. Where appropriate, this theory comes in handy to provide solutions and the alternate actions to be carried out to obtain the intended results for projects to be evaluated. Further, it can be used to enhance decision making and expand conceptions of solutions to any project problems (McClintock, 1990). Rossi, Lipsey and Freeman (2003) describe programme theory as consisting of the organizational plan which deals with how to gather, configure and deploy resources and how to organize programme activities so that the intended service system is developed and maintained.

The theory also deals with the service utilization plan which looks at how the intended target population receives the intended amount of the intended intervention through interaction with the programme service delivery system. Finally, it looks at how the intended intervention for the specified target population brings about the desired social benefit impacts. Uitto (2004) also identifies the advantages of the theory-based framework to M&E to include being able to attribute project outcomes to specific projects or activities and identify the unanticipated and undesired programme or project consequences. Theory-based evaluations enable the evaluator to tell why and how the programme is working (Weiss, 2004). However, this theory is limiting in its approaches as it requires excessive reliance on a collection of data to guide in the evaluation process and this may be costly for projects that are working under tight budgetary allocations. The theory is also limited since it only overemphasizes the impact of the project to the intended people but does not state anything to do with the project executors, whether they have the capacity and ability to collect the data intended for the evaluation.

Summary

The evaluation theory tree developed by Carden and Alkin (2012) summarizes all established conceptual frameworks and approaches advanced by earlier researchers and theorists of evaluation studies. Carden and Alkin (2012) argued that the existing concepts in the evaluation practice are those of the developed world, namely North America, New Zealand, Australia and Europe and as such. The field of evaluation has also been developed based on the needs and perspectives of the developed regions at the expense of low middle-income countries (LMIC). Given the foregoing, conceptual approaches to monitoring and evaluation in the developing countries are either adapted or adopted, which leads to numerous challenges in its implementation in these countries. The ensuing chapter throws light on some established M&E models.

References

Alkin, M. C. & Christie, C. A. (2004). An evaluation theory tree. In: *Evaluation roots: Tracing theorists' views and influences.* SAGE Publications, pp. 12–65.

Berssaneti, F. T. & Carvalho, M. M. (2015). Identification of variables that impact project success in Brazilian companies. *International Journal of Project Management*, 33(3), pp. 638–649, doi:10.1016/j.ijproman.2014.07.002

Cameron, J. (1993). The challenges for monitoring and evaluation in the 1990s. *Project Appraisal*, 8(2), pp. 91–96, doi:10.1080/02688867.1993.9726893

Carden, F. & Alkin, M. C. (2012). Evaluation roots: An international perspective. *Journal of MultiDisciplinary Evaluation*, 8(17), pp. 102–118.

Christie, C. A. & Alkin, M. C. (2013). An evaluation theory tree. In: Alkin, C. M. (ed.). *Evaluation roots: A wider perspective of theorists' views and influences.* Thousand Oaks, CA: SAGE Publications.

Donaldson, L. (2001). Reflections on knowledge and knowledge-intensive firms. *Human Relations*, 54(7), pp. 955–963, doi:10.1177/0018726701547008.Donaldson, 2012

Ika, L. A., Diallo, A. & Thuillier, D. (2012). Critical success factors for World Bank projects: An empirical investigation. *International Journal of Project Management*, 30(1), pp. 105–116, doi:10.1016/j.ijproman.2011.03.005

James, C. (2011). *Theory of change review.* a report commissioned by Comic Relief; http://mande.co.uk/2012/uncategorized/comic-relief-theory-of-change-review/.

Jones, N. et al. (2009). *Improving Impact Evaluation Coordination and Use.* A scoping study commissioned by the DFID Evaluation Department on behalf of NONIE (www.odi.org.uk/resources/download/3177.pdf). Retrieved 25th June 2018.

Kamau, C. G. & Mohamed, H. B. (2015). Efficacy of monitoring and evaluation function in achieving project success in Kenya: A conceptual framework. *Science Journal of Business and Management*, 3(3), p. 82, doi:10.11648/j.sjbm.20150303.14

Katz, I., Newton, B. J., Shona, B. & Raven, M. (2016). Evaluation theories and approaches; relevance for Aboriginal contexts.

Kibebe, L. W. & Mwirigi, P. W. (2014). Selected factors influencing effective implementation of Constituency Development Fund (CDF) projects in Kimilili Constituency, Bungoma County, Kenya. *International Journal of Science and Research*, 3(1), pp. 44–48.

McClintock, C. (1990). Administrators as applied theorists. *New Directions for Program Evaluation*, 1990(47), pp. 19–33, doi:10.1002/ev.1552

McCoy, K. L. Ngari, P. N. & Krumpe, E. E. (2005). *Building monitoring, evaluation and reporting systems for HIV/AIDS programs.* Washington, D.C.: Pact.

Mertens, D. M. & Wilson, A. T. (2012). *Program evaluation theory and practice. A comprehensive guide.* New York: Guildford Press.

Musomba, K. S., Kerongo, F. M., Mutua, N. M. & Kilika, S. (2013). Factors affecting the effectiveness of monitoring and evaluation of constituency development fund projects in Changamwe Constituency, Kenya. *Journal of International Academic Research for Multidisciplinary*, 1(8), pp. 175–216.

Omonyo, A. B. (2015). *Lectures in project monitoring & evaluation for professional practitioners.* Germany: Lambert Academic Publishing.

Papke-Shields, K. E., Beise, C. & Quan, J. (2010). Do project managers practice what they preach, and does it matter to project success? *International Journal of Project Management*, 28(7), pp. 650–662, doi:10.1016/j.ijproman.2009.11.002

Passia, (2006). *Civil Society Empowerment: Monitoring and Evaluation.* Retrieved on 4thJune 2017 from: www.passia.org/seminars/2002/monitoring.htm.

Patton, M. Q. (2008). *Utilization-focused evaluation*, 4th edition. New York: Sage.

Prabhakar, G. P. (2008). Projects and their management: A literature review. *International Journal of Business and Management*, 3(8), pp. 3–9.

Rogers, P. (2008). Using programme theory to evaluate complicated and complex aspects of interventions. *Evaluation*, 14(1), pp. 29–48, doi:10.1177/1356389007084674

Rogers, P. (2014). *Theory of change: Methodological briefs*. Florence: UNICEF Office of Research.

Rossi, P. H., Lipsey, M. W. & Freeman, H. E. (2003). *Evaluation: A systematic approach*, 7th edition. Thousand Oaks. CA: SAGE Publications, Inc.

Shadish, W. R. (1998). Presidential address: Evaluation theory is who we are. *American Journal of Evaluation*, 19, pp. 1–19.

Shadish, W., Cook, T. & Leviton, L. (1991). *Foundations of program evaluation: Theories of Practice*. Newbury Park, CA: Sage.

Shapiro, J. (2004). *Monitoring and Evaluation*. Johannesburg: CIVICUS.

Smith, N. L. (1993). Improving evaluation theory through the empirical study of evaluation practice. *Evaluation Practice*, 14(3), pp. 237–242.

Stein, D. & Valters, C. (2012). *Understanding theory of change in international development*. JSRP Paper No. 1.

Stufflebeam, D. L. & Coryn, C. L. S. (2014). *Evaluation theory, models & applications*, 2nd edition. San Francisco, CA: Jossey-Bass.

Taplin, D. H. & Clark, H. (2012). *Theory of change basics. A Primer on Theory of Change*. New York: ActKnowledge.

Taplin, D. H., Clark, H., Collins, E. & Colby, D. C. (2013). *Theory of change*. New York: Center for Human Environments.

Tengan, C. & Aigbavboa, C. (2016). Evaluating barriers to effective implementation of project monitoring and evaluation in the Ghanaian construction industry. *Procedia Engineering*, 164, pp. 389–394, doi:10.1016/j.proeng.2016.11.635

Uitto, J. (2004). Multi-country cooperation around shared waters: Role of monitoring and evaluation. *Global Environmental Change*, 14, pp. 5–14.

Waithera, L. & Wanyoike, D. M. (2015). Influence of project monitoring and evaluation on performance of youth funded agribusiness projects in Bahati Sub-County, Nakuru, Kenya. *International Journal of Economics, Commerce and Management*, 3(11), pp. 375–394.

Weiss, A. (1995). Human capital vs. signalling explanations of wages. *The Journal of Economic Perspectives*, 9(4), pp. 133–154.

Weiss, C. H. (2004). On theory-based evaluation: Winning friends and influencing people. *The Evaluation Exchange*, 9(4), 2–3.

5 Monitoring and evaluation models

5.1 Abstract

Monitoring and evaluation (M&E) are undertaken for various reasons. The purpose of an M&E will define the approach to be adopted. This chapter presents the evaluation classifications of Vedung (1997), Worthen et al. (2004) and Stufflebeam (2000) and that of the 21st-century evaluation approaches. The chapter informs that Vedung's evaluation classification was influenced by what evaluation is believed to achieve, while Stufflebeam's evaluation classification was influenced by a desire to appraise which 20th-century evaluation approaches were relevant for future use. The 21st-century evaluation approaches are believed to be objective-oriented, management-oriented, consumer-oriented, expertise-oriented, adversary-oriented and participant-oriented evaluation approaches. The chapter argues that evaluators should be integral in the project implementation while using evaluation as a corrective measure in achieving the needed outcome having the broader project outcome in mind.

5.2 Introduction

Omonyo (2015) asserts that evaluation models are points of view or conceptions that various groups of theorists or evaluators are inclined to or approve of. Understanding evaluation models or paradigms is important to generate knowledge on how project M&E should be undertaken to effect the needed or intended outcome on projects. Despite the existence of several evaluation models developed by theorists over the years, the selection of a model for evaluation became a problem and, as such, the need to merge common schools of thought or traditions (Gabriel, 2013). The classification of evaluation models of Vedung (1997), Worthen et al. (2004) and Stufflebeam (2000) is presented below:

5.3 Classification of evaluation models

5.3.1 Evert Vedung's classification

Vedung's classification was influenced by what evaluation is believed to achieve (Gabriel, 2013). His focus was to ensure evaluation satisfied the demand for public service and government. He therefore identified 11 evaluation models and presented his classification as the effectiveness, economic and professional models (Gabriel, 2013). The effectiveness evaluation model concerns evaluation approaches that are instituted by a desire to assess the outcomes of a project, policy or programme. He classifies the following seven evaluation approaches under the effectiveness category: goal-attainment model, side-effect model, goal-free evaluation model, comprehensive evaluation model, client-oriented model and stakeholder model (ibid). The economic models define evaluation approaches that measure the outcomes of public policy or programme relative to the cost incurred like the productivity and efficiency model (ibid). Finally, the professional models focus on the question of who should perform the evaluation (Gabriel, 2013).

5.3.2 Stufflebeam's classification

Gabriel (2013) asserts that Stufflebeam's evaluation classification was influenced by a desire to appraise which of the 20th-century evaluation approaches are relevant for future use and which are not. Stufflebeam classifies 22 evaluation approaches into 4 categories, namely the pseudo-evaluation approach, the question or method-oriented approach, the improvement or accountability approach and the social agenda or advocacy evaluation approach (Gabriel, 2013; Zhang et al., 2011).

5.3.3 Evaluation approaches for the 21st Century

According to Hogan (2010), evaluation approaches relevant and applicable in the 21st century have also been categorized. Worthen et al. (2004) categorized evaluation approaches as objectives-oriented, management-oriented, consumer-oriented, expertise-oriented, adversary-oriented and participant-oriented evaluation approaches (Hogan, 2010).

5.3.4 Stufflebeam's context, input, process and product (CIPP) model

The context, input, process and product (CIPP) model originated in the late 1960s to deliver greater accountability for the US inner-city school district reform project which sought to address the limitations of traditional evaluation approaches (Zhang et al., 2011). The model has over the years been refined and applied in many disciplines, including education. The CIPP evaluation model is recognized as the foremost management-oriented evaluation approach developed by Daniel Stufflebeam (Hogan, 2010). CIPP is an acronym that corresponds to the following core concepts: Context, Input, Process and Product evaluation. According to Mathews and Hudson (2001), context evaluation scrutinizes the programme objectives to determine their social acceptability, cultural relativity and technical adequacy while input evaluation involves an examination of the intended content of the programme. Mathews and Hudson (2001) further opined that process evaluation relates to the implementation of the programme, that is, the degree to which the programme was delivered as planned. Finally, product evaluation is the assessment of programme outcomes (Mathews & Hudson, 2001).

Stufflebeam et al. (2000) noted that the model is intended for the use of service providers, such as policy boards, programme and project staff, directors of a variety of services, accreditation officials, school district superintendents, school principals, teachers, college and university administrators, physicians, military leaders and evaluation specialists. Stufflebeam et al. (2000) further stated that the model is configured for use in internal evaluations conducted by an organization's evaluators, self-evaluations conducted by project teams or individual service providers and contracted or mandated external evaluations. The potential weaknesses of the management-oriented approach as opined by Hogan (2010) may occur from evaluators' giving partiality to top management, from evaluators' occasional inability to respond to questions, from costly evaluation processes and from the assumption that important decisions can be identified in advance.

The CIPP model is a comprehensive framework for guiding formative and summative evaluations of projects, programmes, personnel, products, institutions and systems (Stufflebeam, 2003). Stufflebeam (2003) stated that the CIPP model emphasizes that the evaluation's most important purpose is not to prove but to improve. Evaluation is thus conceived primarily as a functional activity oriented in the long run to stimulating, aiding and abetting efforts to strengthen and improve enterprises (Stufflebeam et al., 2000). However, the model also posits that some programs or other services will prove unworthy of attempts to improve them and should be terminated (Stufflebeam, 2003). By helping stop unneeded, corrupt or hopelessly flawed efforts, evaluations serve an improvement function through assisting organizations to free resources and time for worthy enterprises (Stufflebeam & Shinkfield, 2007).

Consistent with its improvement focus, the CIPP model places a priority on guiding the planning and implementation of development efforts (Stufflebeam et al., 2000). The CIPP model also provides for conducting retrospective summative evaluations to serve a broad range of stakeholders. Potential consumers need summative reports to help assess the quality, cost, utility and competitiveness of products and services they might acquire and use. Other stakeholders might want evidence on what their tax or other types of support yielded (Stufflebeam, 2003). The CIPP model hence brings to the fore the roles of management, that is project leadership, in the implementation of M&E to ensure successful service and product delivery.

5.3.5 Scriven's goal-free evaluation model

The goal-free evaluation model developed by Scriven in 1972 posits that in investigating the set objectives of a project or programme, a broader consideration of other project outcomes should be looked at. Therefore, it is necessary to widely consider beyond the intended outcomes and also look at the unintended outcomes of the project (Omonyo, 2015). While projects are ultimately considered successful when cost, quality and time are achieved, other indirect outcomes such as beneficiary satisfaction and environmental and socio-economic impact of the project should be evaluated as well. These assessments could be done while project implementation is ongoing or at the end or completion of projects while providing meaningful improvement measures. To this end, a logic or programme model is usually developed for the project and tested for validity with the data collected from the evaluation process. This model stresses the approach to monitoring and evaluation particularly regarding data collection and the utilization of the data.

5.3.6 *Stake's responsive evaluation model*

As early as 1975, Stake developed the responsive evaluation model, also referred to as the naturalistic or anthropological model. This approach emphasized the concentration of evaluation on the intended outcomes relating to the programme activities as compared to Scriven's model which sought to place much emphasis on the unintended outcomes of projects. This model argues that the needs of clients are paramount to every project and hence satisfying them should be the main preoccupation of M&E. Gathering project data is key in the M&E process; this notwithstanding, instead of depending on scientific methodologies of experimental psychology, human observations and judgments are heavily relied upon, drawing on a journalistic approach to the evaluation. While relying on qualitative methodologies in a naturalistic evaluation, precise methods for collecting, analyzing and interpreting data are optional.

5.3.7 *Patton's utilization-focused evaluation model*

A management-oriented evaluation model was developed by Patton in 1978. This was referred to as the utilization-focused evaluation model. As has been strongly articulated in earlier sections, M&E serves many purposes, particularly for decision making by the project implementation team to inform ongoing activities (corrective measures) or to inform future projects. Patton argues that decision makers have often ignored evaluation findings; he suggests that as early as possible, in the project planning, key stakeholders such as relevant decision makers and the audience of evaluation reports who utilize evaluation findings must be identified. Establishing effective collaboration between the evaluation team and the consumers of the evaluation findings is therefore important.

5.3.8 *Guba's ethnographic evaluation model*

In 1978, Guba proposed the ethnographic evaluation model. As argued by this model, evaluators of projects are an integral part to the project implementation as they are involved in the project from the inception up until the completion of the project and participate in the day-to-day monitoring and supervision of the project. The philosophy behind Guba's model is to afford evaluators the opportunity to obtain a detailed description of the project being implemented and convey the same to the project stakeholders. The model advocates for the involvement and communication flow among the key stakeholders of the projects to get involved in the M&E during project implementation.

Summary

Chapter 5 provides an understanding of the evaluation models and their underpinnings. The classification of evaluation models by Vedung (1997), Worthen et, al. (2004) and Stufflebeam (2001) were presented and discussed. Similarly, the evaluation approaches of the 21st century were categorized as objective-, management-, consumer-, expertise-, adversary- and participant-oriented. Stufflebeam's context, inputs, process and product (CIPP) model, Scriven's goal-free evaluation model, Stake's responsive evaluation model, Patton's utilization-focused evaluation model and Guba's ethnographic evaluation model were also discussed. The next chapter provides a conceptual understanding of M&E in construction.

References

Gabriel, K. (2013). *A conceptual model for a programme monitoring and evaluation information system*. Stellenbosch: Stellenbosch University.

Hogan, R. L. (2010). The historical development of program evaluation: Exploring past and present. *Online Journal for Workforce Education and Development*, 2(4), p. 5.

Mathews, J. M. & Hudson, A. M. (2001). Guidelines for evaluating parent training programs. *Family Relations*, 50(1), 77–87.

Omonyo, A. B. (2015). *Lectures in project monitoring & evaluation for professional practitioners*. Germany: Lambert Academic Publishing.

Stufflebeam, D. L. (2003). The CIPP model for evaluation. In: *International handbook of educational evaluation*, pp. 31–62. Dordrecht: Springer.

Stufflebeam, D. L. & Shinkfield, A. J. (2007). *Evaluation theory, models, and applications*. San Francisco, CA: Jossey-Bass.

Stufflebeam, D. L., Madaus, G. F. & Kellaghan, T. (eds.). (2000). *Evaluation models: Viewpoints on educational and human services evaluation* (2nd edn.). Netherlands: Springer Netherlands.

Vedung, Evert, 1997, *Public Policy and Program Evaluation*, 209–245, Piscataway, NJ and London: Transaction.

Worthen, B.R., Sanders, J. R. & Fitzpatrick, J. L. (2004). *Educational evaluation: Alternative approaches and practical guidelines* (3rd edn.). Boston: Allyn & Bacon.

Zhang, G., Zeller, N., Griffith, R., Metcalf, D., Williams, J., Shea, C. & Misulis, K. (2011). Using the context, input, process, and product evaluation model (CIPP) as a comprehensive framework to guide the planning, implementation, and assessment of service-learning programs. *Journal of Higher Education Outreach and Engagement*, 15(4), pp. 57–84.

6 Conceptual IME model for construction project delivery

6.1 Abstract

The chapter conceptualizes monitoring and evaluation as a five-factor model. The chapter asserts that the M&E of projects will require the involvement of all parties to the project (stakeholder), sufficient budgetary allocation, technical capacity and training, effective leadership and communication. An empirical literature review approach was adopted and the frequency of reviewed factors indicated the importance of the factor in facilitating effective M&E of the construction process. The review informs that the five-factor conceptualized IME model potentially will result in the desired project outcomes: timely completion of projects, conformity to specification, achieving project cost, stakeholder satisfaction, health & safety, value for money, environmental performance and end user and client satisfaction. The knowledge gained will help in the understanding of the elements and components of the M&E system relating or influencing each other to facilitate the achievement of project targets and goals.

6.2 Introduction

Several factors influence the effective implementation of monitoring and evaluation. These factors are broadly seen as inputs and processes. Musomba et al. (2013) studied the factors affecting the effectiveness of M&E of constituency development fund projects in Kenya which established four independent variables that influence the effective M&E to affect project success. The study indicates that the level of training, budgetary allocation and stakeholder participation as well as political influence had a greater chance to ensure good M&E. In a similar study, the efficacy of M&E functions in achieving project success in Kenya was investigated by Kamau and Mohamed (2015). The study broadly categorized the factors that influenced M&E into three. First, the study talks about the strength of the monitoring team and the environment that will ensure effective M&E of projects such as funds' availability, use of technology in M&E, stakeholder representation and the frequency of M&E. The second category is the approach to M&E which also describes the tools that will be used to ensure M&E are effective. Tools such as the basic research, internal audits, balanced scorecard and the log frame matrix are discussed. Finally, the influence of politics is acknowledged as affecting the effective implementation of M&E (Kamau & Mohamed, 2015).

Other factors such as human resources, the project organizational culture, stakeholders and advocacy substantially influenced the effective M&E system of public health programmes studied among school-based handwashing programmes in Kwale County, Kenya (Otieno Okello, 2015). Stakeholder participation or involvement as well as political influence were identified as influential factors in the success of M&E (Mwangi, Nyang'wara & Ole Kulet, 2015; Waithera & Wanyoike, 2015). Mwangi et al. (2015) further identified the technical capacity of the M&E team and budgetary allocation as important influencing factors to the practice of M&E. The level of training of M&E staff and continuous training of staff on M&E were also identified as necessary factors (Musomba et al., 2013; Waithera & Wanyoike, 2015). In addition, Hardlife and Zhou (2013) found that the availability of resources such as time, sufficient finances, adequate skilled personnel, technical competence regarding the application and utilization of an M&E system and a favourable administrative culture significantly influence the success of an M&E system.

Mugo and Oleche (2015) studied the M&E of development projects and economic growth in Kenya and identified the training of the personnel on M&E, stakeholders' participation, institutional guidelines and the amount of budgetary allocation as noteworthy factors determining the successful implementation of an M&E system in development projects in Kenya. A study conducted in Ghana identified barriers to the implementation of M&E such as weak institutional capacity, lack of resource and budgetary allocation and the weak institutional approach to M&E. These were factors that militate against the effective implementation of M&E of projects in the Ghanaian construction industry (Tengan & Aigbavboa, 2016).

In other studies, Crawford and Bryce (2003) acknowledged that for M&E to be effective to mitigate poor project performance, demonstrate accountability and promote organizational learning, project M&E information systems (ISs) are imperative. Kimweli (2013), studying the role of M&E practices in the success of donor-funded food security intervention projects in Kibwezi District, Kenya, bemoaned the failure of M&E toward achieving food security to the exclusion of the community in the M&E process. Again in Nairobi County, Kenya, a study to determine the effectiveness of the M&E of government-funded water projects revealed that budgetary allocation, monitoring team capacity, stakeholder involvement and management skill influence the successful completion of donor water projects owing to inadequate M&E (Ogolla & Moronge, 2016).

Seasons (2003) also studied the realities of M&E in municipal planning. In his study, six factors were outlined as determinants of the effectiveness of M&E practice. Limited resource allocation for M&E was mentioned. He referred to the need for enough time to undertake the effective evaluation, adequate funds for the M&E and the required competence and expertise. The planning process influenced the effectiveness of M&E. He asserted that it appears to be the case that M&E are usually forgotten during the planning stage of the project or programme implementation. That is, M&E are not planned for before projects or programmes get underway, resulting in dire consequences for the project outcome. Appropriate indicators served as the benchmark to measure performance, the ultimate of the M&E process. In the absence of appropriate indicators, M&E are ineffective. The political realities influence M&E. The intention of conducting M&E is paramount to its success. The study revealed that M&E were carried out merely for political exigencies and not for the effectiveness and efficiency of the goal or objective of the project or programme. The important determinant is the causality or relationship between goals and outcome. Difficulty in establishing the link between the inputs and outcomes due to unplanned factors such as changes in market conditions and political decisions rendered the M&E process ineffective and, finally, the organization culture towards M&E influenced its success.

Mugambi and Kanda (2013) carried out a study to determine factors affecting the M&E of community-based projects. The study identified the relationship or involvement of stakeholders, budgetary allocation, the M&E process and communication. Mulandi (2013) studied the performance of M&E systems in selected non-governmental organizations in Kenya. The study revealed data quality, human capacity, the use of the logical framework and utilization of M&E findings as factors that influence the performance of M&E systems regarding access and the provision of accurate and quality information. Moderating variables, stakeholder participation and funding as well as extraneous variables and donor priority are required to ensure that the above-mentioned factors caused the required performance of M&E systems (Mulandi, 2013). Similarly, Muiga (2015) studied influencing factors regarding the use of M&E systems in public projects where it was found that training levels, budgetary allocation, stakeholder participation and political interference induced the M&E systems in public projects. Before the study by Muiga (2015), Oloo (2011) identified the level of training, institutional framework, budgetary allocation and stakeholder participation as important factors influencing the effectiveness of M&E of constituency development fund projects in the Likoni constituency with a politically influenced environment as a moderating variable.

6.3 Key determinants of effective monitoring and evaluation

Table 6.1 presents the summary of the determinants of effective M&E. The frequency of the determinants as cited in the reviewed literature was indicated so as to establish the ranking of determinants that influence effective implementation of M&E. In sum, the top three most discussed influencing factors of effective M&E established through the review of the relevant literature are budgetary allocation, stakeholder involvement and technical capacity and training from a combination of diverse industries and context. Similarly, two other factors which appear significant but have not been thoroughly studied for their level of influence on construction project M&E are communication and leadership of M&E implementation. Together, these five M&E-determining factors are considered for the development of the integrated conceptual model.

Table 6.1 Determinants of effective M&E

Sl. No.	Monitoring and evaluation determinants	A	B	C	D	E	F	G	H	I	J	K	L	M	N	O	P	Freq of Occurrence	Prioritized rank
1	Approach to M&E	✓	✓					✓	✓								✓	5	4th
2	Political influence on M&E	✓	✓			✓		✓	✓									5	4th
3	M&E advocacy				✓													1	8th
4	Project organizational culture on M&E				✓			✓										2	7th
5	Stakeholders' involvement in M&E	✓		✓	✓	✓	✓	✓				✓	✓	✓	✓		✓	10	1st
6	Technical capacity and training of the M&E team				✓	✓									✓	✓		4	6th
7	Budgetary allocation for M&E		✓	✓	✓	✓	✓	✓				✓	✓	✓			✓	9	2nd
8	Institutional guideline for M&E		✓	✓	✓	✓	✓	✓				✓	✓	✓				9	2nd
9	Institutional capacity for M&E							✓										1	8th
10	Communication of M&E findings								✓									1	8th
11	Appropriate indicators for M&E								✓									1	8th
12	Causality or relationship between M&E goals and outcome							✓										1	8th
13	Monitoring and evaluation information systems (MEIS)										✓							1	8th
14	Community participation in the M&E process											✓						1	8th
15	Management skill in M&E												✓					1	8th
16	M&E data quality														✓			1	8th
17	Use of M&E logical framework															✓		1	8th
18	Organizational leadership for M&E																✓	1	7th

Source: Author's literature review

A = Kamau and Mohamed (2015), **B** = Otieno Okello (2015), **C** = Waithera and Wanyoike (2015), **D** = Mwangi et al. (2015), **E** = Musomba et al. (2013), **F** = Mugo and Oleche (2015), **G** = Tengan and Aigbavboa (2016), **H** = Mugambi & Kanda (2013), **I** = Seasons (2003), **J** = Crawford and Bryce (2003), **K** = Kimweli (2013), **L** = Ogolla & Moronge (2016), **M** = Muiga (2015), **N** = Oloo (2011), **O** = Mulandi (2013), **P** = Njama (2015)

6.3.1 Stakeholder involvement

Stakeholders had evolved since 1963 when it first appeared at the Stanford Research Institute (SFI) in the literature of organizational management till date (Elias & Cavana, 2000). In M&E, the involvement of stakeholders is critical for many reasons. Key project stakeholders' involvement in M&E will drive the need to meet their expectations and to create an opportunity to share M&E responsibilities. Involvement of stakeholders in any organization activities is crucial for its survival (Freeman, 2004; Freeman et al., 2010; Hörisch, Freeman & Schaltegger, 2014), hence their involvement in the M&E process is necessary and cannot be overlooked. Freeman et al. (2010) described stakeholders as a group of people in whose support lies the survival of the organization or company. In modern literature, the composition of stakeholders has evolved to include individuals or group of people who are directly or indirectly involved, who influence or are being influenced by the outcome of a project (Bourne, 2010; Elias & Cavana, 2000; Eyiah-Botwe, Aigbavboa & Thwala, 2016; Hermans, Haarmann & Dagevos, 2011). In construction project delivery, several stakeholders exist to participate in project implementation. However, their involvement in the M&E process is limited.

For the effective M&E of construction projects at the local government level, Elias and Cavana (2000) argue for the identification of project stakeholders based on their financial support of the project. In Ghana, the central government, being the primary financier of development in the country, has the right to ensure accountability of its financial commitment towards project delivery, just as all donor agencies or sponsors of projects. The municipal, metropolitan and district assemblies (MMDAs) at the local government level represent this interest of government (local government) and thus ensure that projects are tracked for progress as well as the efficient, effective sustainability of the project. Project consultants, according to the Project Management Institute (PMI) (2010), are also organizations with the requisite knowledge, skills and experience in a particular area of discipline who assist organizations to improve project practice and management. They are therefore external organizations that are entrusted with the implementation of the project and receive a fee for such services. Implementers of the project or development, i.e. the contractors, cannot be left out. Contractors are central to the M&E process (Mwangu & Iravo, 2015) since it is their performance that is checked against the desired standards and specifications. Their commitment to construct per specification will contribute to the successful project delivery.

The end users or beneficiaries of the project also need to be involved in the project through M&E. Beneficiary involvement in the M&E process increases their interest in the project and opportunities such as job opportunities are created for them. Suppliers and service providers, the Ghana Water Company Limited (GWCL) and the Electricity Company of Ghana (ECG) are all essential stakeholders in the construction project delivery and their involvement in the M&E of projects may be necessary. Stakeholders are therefore necessary and the project cannot succeed without them (Muiga, 2015). Stakeholders must therefore be identified early prior to the start of the project (Elias & Cavana, 2000). However, caution is necessary about the number of stakeholders to be involved in the M&E since greater numbers can unduly influence the smooth implementation of the project and the evaluation process (Patton, 2010). It is significant to ensure that stakeholders participate in the M&E process through representation since all the interested parties cannot be brought on board to participate in the process (Hermans et al., 2011) and also considering the scarce resources available for M&E in most cases. Indeed, a good monitoring team is said to be that which has a good stakeholder representation (Kamau & Mohamed, 2015).

Stakeholder involvement in this book is premised on the effective and conducive project environment to foster effective stakeholder involvement and interaction in the entire M&E process. The recognition and composition of stakeholders in project committees and having regard for their competencies, knowledge and interest are important to encourage and sustain the involvement of interested parties in the M&E practice (Magondu, 2013). It is important to explicitly outline the stake or interest of all identified stakeholders (Elias & Cavana, 2000). The level of involvement in the project M&E is greatly influenced by the kind of stakeholders; contractors and consultants are integral (key parties) in the project implementation and as such, are usually engaged in the M&E, unlike stakeholders such as the beneficiary community who are only involved in the project during community entry and project closure. Key to stakeholder involvement is the efficient management of the stakeholder relationship, power structure and influence in the project (Naidoo, 2011). Regardless of the kind of stakeholder, the level of motivation as well as training can drive the active involvement of all stakeholders (Ruwa, 2016). The process of M&E results in the generation of an M&E report; proper communication of M&E reports among stakeholders and appropriate implementation of these reports are imperative to sustain the continuous involvement of stakeholders in M&E. Their constant commitment is sustained through the provision of training and development on the importance of, use and need for M&E results.

6.3.2 *Budgetary allocation*

Allocating adequate financial resources for M&E during budgeting is imperative to achieve the effectiveness of M&E (Kimani, 2014). The successful implementation of M&E is firmly rooted in the provision of the adequate financial resource (Kimani, 2014; Mugambi & Kanda, 2013; Mugo et al., 2015; Muiga, 2015; Musomba et al., 2013; Ogolla & Moronge, 2016; Oloo, 2011; Seasons, 2003). Sufficient funds allocated for monitoring and evaluation activities are necessary. Hwang and Lim (2013) are of the view that budgetary performance could lead to project success. It is therefore vital to realistically draw up a clear budget line specific to M&E and incorporate it into the overall project cost. Studies have indicated that the method of budgeting, i.e. a fixed percentage of the contract price or fixed amount for M&E, is significant. A minimum recommendation of between 3% and 5% of the total project cost has been advocated as a reasonable budget allocation for

M&E. The scope and complexity of M&E, as well as the number of stakeholders involved in the M&E, should, however, be considered when budgeting since it will influence the amount of budget allocated. A form of budgetary disbursement for M&E activities is necessary to ensure that funds are available throughout the M&E process. Timely release of M&E funds having regard to the M&E duration will also influence the allocation of budget for M&E. The sources of funding (internally generated funds or donor funding) is critical in sustaining budgetary allocation. To guarantee that budgeting is done correctly and efficiently, the need for periodic auditing (internal/external) of the M&E budget will ensure budget allocations are sustained and rightly so, influence the M&E of projects effectively.

6.3.3 Technical capacity and training

M&E is a technical activity and therefore the technical capacities of an M&E unit and its staff/team are important as these underpin the effective implementation of M&E. Technical capacity is a unique and practical knowledge possessed by the project team whereas training is a planned experience that assists individuals to acquire new skills, knowledge and attitude to address developmental problems (USAID, 2010). The strength of an organization in dealing with M&E is associated with its human resource capacity (Muiga, 2015); without skilled personnel, M&E systems cannot work on their own (Mulandi, 2013). The United Nations Development Programme (UNDP) (2009) indicated that capacity development is the process of gaining, strengthening and maintaining skills and capabilities for achieving developmental goals and objectives within the specific time frame. The integration of the conventional training approach is explained by the UNDP to describe training on the use and application of new technology available to a capacity development approach which sees training as part of a comprehensive programme seeking to address capacity issues; learning to use and apply readily available technologies which are best suited for the industry; personal development of employees with an incentive for innovation; empowering the development of trainers, trainees and associating personal performance to team performance which are critical for development (UNDP, 2009).

Studies have acknowledged the important role played by the level of the technical capacity and training in M&E to achieve project success (Mugo et al., 2015; Muiga, 2015; Mulandi, 2013; Musomba et al., 2013; Ogolla & Moronge, 2016; Oloo, 2011; Otieno Okello, 2015; Waithera et al., 2015). The presence of systems and financial resources are necessary but on their own, they cannot sufficiently guarantee project success, but rather ownership and technical capacity development (UNDP, 2009). The question to ask is whether the M&E team has the necessary capacity and strength to undertake effective M&E. The UNDP (2009) further indicates that effective capacity development is better achieved in an enabling environment at the organizational level and by individuals. An enabling environment concerns the organizational policies, rules and power relationships within an organization which is to function efficiently with the involvement of individuals (stakeholders) who have acquired the required skills, experience and knowledge through prior educational training or observation and doing (involvement). These have been summed up into four enablers by the UNDP as an institutional arrangement, leadership, accountability and knowledge (UNDP, 2009).

The Paris Declaration on Aid Effectiveness adopted in 2005 and re-affirmed in 2008 in Accra-Ghana was signed by over one hundred donors and developing countries (Bissio, 2007). According to Otoo, Agapitova & Behrens, (2009), the capacity to plan, manage, implement and account for results ensured the achievement of development goals and objectives, hence the need for development of M&E capacity. Planning, managing and implementing the M&E process of projects are successful with the right training in M&E. Otoo et al. (2009) indicate that factors presented in the framework can be used as a basis to measure the impact of capacity development. The outlined indicators include conduciveness of the sociopolitical environment, the efficiency of policy instrument and the effectiveness of the organizational arrangement. A study in Indonesia revealed that capacity development is necessary to combat challenges impeding the achievement of success (Subijanto, Ruritan & Hidayat, 2013). The study further indicated the key success factors for capacity development as strong leadership, incentive schemes and the spirit of innovation, willingness and eagerness to take on new challenges.

Given the influence of capacity development on the effective and efficient M&E implementation of projects, the training, either self or corporate training of staff, of an M&E unit cannot be overstated. Kontoghiorghes (2001) informs that for training to occasion the needed capacity development, the effectiveness of the training, the level of trainee knowledge prior to the immediate training, management and organizational support, the frequency of training and the degree of employee involvement in training are critical. Similarly, Punia and Kant (2013) reviewed the factors affecting training effectiveness and identified management support and the style or form of training as key in developing capacity. In a meta-analysis study of the effectiveness of deception detection training, Driskell (2012) identified the content of training and trainee expertise as moderators of training effectiveness.

6.4 Proposed IME conceptual model

A conceptual framework, also referred to as a research framework (Frankel & Gage, 2007), presents the relationships between relevant factors that may have an impact on the achievement of goals and objectives. Frankel and Gage (2007) describe a conceptual framework as useful for identifying and demonstrating the factors and relationships that influence the outcome of a project. Ogolla and Moronge (2016) also defined a conceptual framework as a virtual or written product that describes, either graphically or in description form, the main elements, the key factors, concepts or variables and the presumed relationships that exist among them.

The proposed conceptual framework therefore presents the degree of influence and relationship that exists between factors, i.e. variables which will result in a desirable outcome or impact. The variables defined here represent the independent and dependent variable. The independent variable influences and determines the effect of dependent variables, namely stakeholder involvement, budgetary allocation, political influence, technical capacity and training, and change management and organizational learning. The dependent variable is that factor which is observed and measured to determine the effect of the independent variable. The integrated conceptual framework below is therefore proposed to ensure effective M&E in construction project delivery.

6.5 Theoretical underpinning of the conceptual model

This section of the theoretical framework focuses on the significant reasons underlying the theory chosen for the study. While the works on M&E have not generated consensus on a single theory to underpin effective M&E implementation, the programme theory and the theory of change summarize the essential variables needed for effective M&E; hence, they both merit as justifiable theories to underpin the current study. The programme theory evaluates how well a programme or project is designed to achieve its intended outcomes. This helps put an emphasis on areas that need critical attention for effective M&E. Similarly, for M&E to be effectively implemented, an understanding of what needs to be monitored and evaluated is a critical variable that demands attention. This will therefore require the development of the capacities of M&E staff and teams of an organization through training to fulfil the set goals of the project. The theory of change rightly described the capacity development of M&E teams as articulating the underlying expectation of the M&E process. Another critical matter needing urgent attention for effective M&E is to ensure accountability of the M&E process. Also, a key aspect of the theory of change is the opportunity for organizations to communicate selected changes in processes to partners. Taplin, Clark, Collins and Colby (2013) informed that the theory of change drives communication through the outcome pathways and narrative to stakeholders and builds core capacities for M&E. The combined theories underscore the variables' stakeholder involvement, budgetary allocation, technical capacity and training, M&E communication and leadership conceived for the development of an IME model for construction project delivery in Ghana.

6.6 Structural components and specification of the conceptual IME model

This book conceptualizes IME model for construction project delivery founded on the relationship between the exogenous variables described above and the other causal factors which connect both the objective and the subjective measurements. The variables were identified from the review of the existing literature and confirmed through a Delphi study. This brought to light critical determinants of the factors that ensure effective M&E of construction project delivery. The M&E of construction project delivery is influenced by the involvement of stakeholders' features (SIF), budgetary allocation features (BAF), technical capacity and training features (TC&TF), M&E leadership features (M&ELF) and M&E communication features (M&ECF) as represented in Figure 6.1.

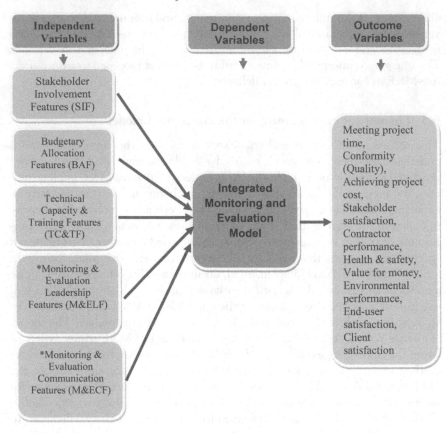

Figure 6.1 Integrated M&E conceptual model for construction project delivery.

The 5 primary constructs for the developed conceptual IME model revealed 15 measurement variables for stakeholder involvement features (SIF), 12 measurement variables for budgetary allocation features (BAF), and 10 measurement variables for technical capacity and training features (TC&TF). M&E leadership features (M&ELF) had 14 measurement variables while 12 measurement variables were identified for M&E communication features (M&ECF). M&E leadership and communication were the 2 identified knowledge gaps in M&E research, hence their introduction in the IME model to ensure effective M&E of construction project delivery. Table 6.2 below illustrates the measurement variables for each of the five identified latent variables:

6.7 IME model justification

While research on construction project M&E has not generated consensus on a single theory to underpin the effective implementation of M&E, the programme theory and the theory of change encapsulate the essential variables needed for effective M&E, hence they both merit as justifiable theories for the conceptualized

Table 6.2 Latent variables and measuring constructs of the conceptual IME model

Latent variable constructs	Measurement variables
Stakeholder Involvement Features (SIF)	1 Engaging stakeholders in M&E
	2 Providing stakeholder need for M&E
	3 Recognition of project stakeholders
	4 Motivating stakeholder towards M&E
	5 Experience of stakeholders in M&E
	6 Stakeholder interest in M&E
	7 Stakeholder expectation in M&E
	8 Identifying stakeholders
	9 Managing stakeholders' power structures
	10 Influence of stakeholders on the project
	11 Stakeholders' involvement in M&E
	12 Collaboration at all levels among stakeholders
	13 Satisfying stakeholders
	14 Training stakeholders on M&E
	15 Developing stakeholder capacity on M&E
Budgetary Allocation Features (BAF)	1 Sufficient budget allocated for M&E
	2 Availability of funds for the M&E activities
	3 The clear budget line for M&E
	4 Method of budgeting for M&E
	5 Form of M&E budget audit
	6 The frequency of M&E audit (internal and external controls)
	7 Scope of M&E
	8 Complexity of M&E
	9 Duration of M&E
	10 Source of funding for M&E
	11 M&E budget record-keeping
	12 Budgetary disbursement for M&E
Technical Capacity and Training Features (TC&TF)	1 The frequency of the M&E training
	2 The content of the M&E training
	3 The planning process of the M&E training
	4 The education level of M the &E staff
	5 Level of the M&E training
	6 Form of the M&E training
	7 Staff expectation before the M&E training
	8 Staff level of involvement in M&E training
	9 Staff's desire for the M&E training
	10 Management support for the M&E training
Monitoring and Evaluation Leadership Features (M&ELF)	1 The leadership style of the M&E team leader
	2 The attitude of the M&E team leader
	3 The vision of the M&E team leader
	4 The commitment of the M&E team leader
	5 Traits (personality and behavioral) of the M&E team leader
	6 Managerial skills of the M&E team leader
	7 Gender of the M&E team leader
	8 Competencies of the M&E team leader
	9 The organizational environment for M&E
	10 Project environment for M&E
	11 Knowledge level of the M&E team leader
	12 Performance of the M&E team leader
	13 Communication style of the M&E team leader
	14 Interpersonal skills of the M&E team leader

(*Continued*)

Table 6.2 Latent variables and measuring constructs of the conceptual IME model (*Continued*)

Latent variable constructs	Measurement variables
Monitoring and Evaluation Communication Features (M&ECF)	1 Channel for communicating the M&E findings 2 Reducing the distortion in the communication of the M&E findings 3 Communicator (Sender) of the M&E findings 4 The intended audience of the M&E findings 5 The relevance of the M&E findings 6 Effective communication skills amongst the M&E team 7 Appropriate feedback channel for the M&E findings 8 Access to the M&E information 9 Proper reporting for M&E 10 Proper communication structure for M&E 11 Standardization of M&E documents 12 Effective listening skills

IME model. The programme theory evaluates how well a programme or project is designed to achieve its intended outcomes. This helps emphasize on areas that need critical attention for effective M&E. Similarly, for M&E to be effectively undertaken, an understanding of what needs to be monitored and evaluated is a critical variable that demands attention. This will therefore require the development of the capacities of M&E staff and teams of an organization through training to achieve the relevant M&E competencies to achieve the object of the project. The theory of change rightly described the capacity development of M&E teams as articulating the underlying expectation of the M&E process. Another critical matter needing urgent attention for effective M&E is to ensure accountability of the M&E process. Also, a key aspect of the theory of change is the opportunity for organizations to communicate selected change processes to partners. Taplin et al. (2013) informed that the theory of change drives communication through the outcome pathways and narrative to stakeholders and builds core capacities for M&E. The combined theories underscore the variables' stakeholder involvement, budgetary allocation, technical capacity and training, M&E communication and leadership conceived for the development of an IME model for construction project delivery.

Summary

The chapter discussed the conceptual integrated monitoring and evaluation model for construction project delivery. Five exogenous variables constituting the IME model are stakeholder involvement, budgetary allocation, technical capacity and training, and M&E communication and M&E leadership. The study relied on both the theory of change and the programme theory to justify the philosophy behind the conceptual model developed for the study. The book posits that M&E are effective to bring about the desired change in the project when the relevant stakeholders are involved, sufficient budget is allocated, the technical capacities of organization staffs are developed and effective communication and effective leadership systems are provided in the monitoring and evaluation process. The next chapter discusses the gaps identified in the literature of construction M&E practice and that need to be evaluated to develop a holistic integrated M&E model.

References

Bissio, R. (2007). Paris Declaration on Aid Effectiveness: Application of the Criteria for Periodic Evaluation of Global Development partnerships — As Defined in Millennium Development Goal 8 — From the Right to Development Perspective: the Paris Declaration on Aid Effectiveness(Geneva: Human Rights Council).

Bourne, L. (2010). Why is stakeholder management so difficult? In: *Congresso International*. Colombia: Universidad Ean Bogota.

Crawford, P. & Bryce, P. (2003). Project monitoring and evaluation: A method for enhancing the efficiency and effectiveness of aid project implementation. *International Journal of Project Management*, 21(5), pp. 363–373, doi:10.1016/S0263-7863(02)00060-1

Driskell, J. E. (2012). Effectiveness of deception detection training: A meta-analysis. *Psychology, Crime & Law*, 18(8), pp. 713–731, doi:10.1080/1068316X.2010.535820

Elias, A. A. & Cavana, R. Y. (2000). Stakeholder analysis for systems thinking and modeling. In: *Victoria University of Wellington, New Zealand. Conference Paper*.

Eyiah-Botwe, E., Aigbavboa, C. & Thwala, W. D. (2016). Mega construction projects: Using stakeholder management for enhanced sustainable construction. *American Journal of Engineering Research*, 5(5), pp. 80–86.

Frankel, N. & Gage, A. (2007). *M&E fundamentals: A self-guided minicourse*. MEASURE evaluation, Inter-agency Gender Working Group, Washington, D.C.: U.S. Agency for International Development.

Freeman, R. E. (2004). The stakeholder approach revisited. *Zeitschrift für Wirtschafts-und Unternehmensethik*, 5(3), pp. 228–241.

Freeman, R. E., Harrison, J. S., Wicks, A. C., Parmar, B. L. & De Colle, S. (2010). *Stakeholder theory: The state of the art*. Cambridge, U.K.: Cambridge University Press.

Hardlife, Z. & Zhou, G. (2013). Utilisation of Monitoring and Evaluation Systems by Development Agencies: The Case of the UNDP in Zimbabwe. *American International Journal of Contemporary Research*, 3(3), pp. 70–83.

Hermans, F. L. P., Haarmann, W. M. F. & Dagevos, J. F. L. M. M. (2011). Evaluation of stakeholder participation in monitoring regional sustainable development. *Regional Environmental Change*, 11(4), pp. 805–815, doi:10.1007/s10113-011-0216-y

Hörisch, J., Freeman, R. E. & Schaltegger, S. (2014). Applying stakeholder theory in sustainability management: Links, similarities, dissimilarities, and a conceptual framework. *Organization & Environment*, 27(4), pp. 328–346.

Hwang, B. G. & Lim, E. S. J. (2013). Critical success factors for key project players and objectives: Case study of Singapore. *Journal of Construction Engineering and Management*, 139(2), pp. 204–215.

Kamau, C. G. & Mohamed, H. B. (2015). Efficacy of monitoring and evaluation function in achieving project success in Kenya: A conceptual framework. *Science Journal of Business and Management*, 3(3), p. 82, doi:10.11648/j.sjbm.20150303.14

Kimani, R. N. (2014). *The effect of budgetary control on effectiveness of non-government organisations in Kenya*. Kenya: School of Business, University of Nairobi.

Kimweli, J. M. (2013). The role of monitoring and evaluation practices to the success of donor funded food security intervention projects: A case study of Kibwezi District. *International Journal of Academic Research in Business and Social Sciences*, 3(6), p. 9.

Kontoghiorghes, C. (2001). Factors affecting training effectiveness in the context of the introduction of new technology – A US case study. *International Journal of Training and Development*, 5(4), pp. 248–260, doi:10.1111/1468-2419.00137

Magondu, A. (2013). *Factors influencing implementation of monitoring and evaluation in HIV research projects: A case of Kenya AIDS Vaccine Initiative (KAVI)*. Thesis. Kenya: University of Nairobi.

Mugambi, F. & Kanda, E. (2013). Determinants of effective monitoring and evaluation of strategy implementation of community-based projects. *International Journal of Innovative Research and Development*, 2(11) pp 67–73

Mugo, P. M. & Oleche, M. O. (2015). Monitoring and evaluation of development projects and economic growth in Kenya. *International Journal of Novel Research in Humanity and Social Sciences*, 2(6), pp. 52–63.

Muiga, M. I. J. (2015). *Factors influencing the use of monitoring and evaluation systems of public projects in Nakuru County*. Kenya: University of Nairobi.

Mulandi, N. M. (2013). *Factors influencing performance of monitoring and evaluation systems of Non-governmental organizations in governance: A case of Nairobi, Kenya*. University of Nairobi.

Musomba, K. S., Kerongo, F. M., Mutua, N. M. & Kilika, S. (2013). Factors affecting the effectiveness of monitoring and evaluation of constituency development fund projects in Changamwe Constituency, Kenya. *Journal of International Academic Research for Multidisciplinary*, 1(8), pp. 175–216.

Mwangi, J. K., Nyang'wara, B. M. & Ole Kulet, J. L. (2015). Factors affecting the effectiveness of monitoring and evaluation of Constituency Development Fund projects in Kenya: A case of Laikipia West Constituency. *Journal of Economics and Finance*, 6(1), pp. 74–87.

Mwangu, A. W. & Iravo, M. A. (2015). How monitoring and evaluation affects the outcome of Constituency Development Fund projects in Kenya: A case study of projects in Gatanga Constituency. *International Journal of Academic Research in Business and Social Sciences*, 5(3), doi:10.6007/IJARBSS/v5-i3/1491.

Naidoo, I. A. (2011). *Governance in South Africa: A case study of the Department of Social Development*. PhD Thesis. Johannesburg: University of Witwatersrand.

Ogolla, F. & Moronge, M. (2016). Determinants of effective monitoring and evaluation of government funded water projects in Kenya: A case of Nairobi County. *Strategic Journal of Business & Change Management*, 3(1).329–358

Oloo, D. O. (2011). *Factors affecting the effectiveness of monitoring and evaluation of constituency development fund projects in Likoni Constituency, Kenya*. Kenya: University of Nairobi.

Otieno Okello, L. (2015). Determinants of effective monitoring and evaluation system of public health programs: A case study of school-based hand washing program in Kwale County, Kenya. *International Journal of Economics, Finance and Management Sciences*, 3(3), p. 235, doi:10.11648/j.ijefm.20150303.20

Otoo, S., Agapitova, N. & Behrens, J. (2009). *The capacity development results framework: A strategic and results-oriented approach to learning for capacity development*. Washington, D.C., U.S.A.: The World Bank Institute.

Patton, M. Q. (2010). *Developmental evaluation: Applying complexity concepts to enhance innovation and use*. New York: The Guilford Press.

Project Management Institute (PMI). (2010). *A guide on how to select a project management consultant.* UK: PMI.

Punia, B. & Kant, S. (2013). A review of factors affecting training effectiveness vis-a-vis managerial implications and future research directions. *International Journal of Advanced Research in Management and Social Sciences,* 2(1).151–164

Ruwa, M. C. (2016). *The influence of stakeholder participation on the performance of donor funded projects: A case of Kinango Integrated Food Security and Livelihood Project (KIFSLP), Kwale County, Kenya.* Master's dissertation. Kenya: University of Nairobi.

Seasons, M. (2003). Monitoring and evaluation in municipal planning: Considering the Realities. *Journal of the American Planning Association,* 69(4)430–440

Subijanto, T. W., Ruritan, H., Raymond Valiant & Hidayat, F. (2013). Key success factors for capacity development in the Brantas River Basin organisations in Indonesia. *Water Policy,* 15(S2), p. 183, doi:10.2166/wp.2013.019

Taplin, D. H., Clark, H., Collins, E. & Colby, D. C. (2013). *Theory of change.* New York: Center for Human Environments. Technical Papers.

Tengan, C. & Aigbavboa, C. (2016). Evaluating barriers to effective implementation of project monitoring and evaluation in the Ghanaian construction industry. *Procedia Engineering,* 164, pp. 389–394, doi: 10.1016/j.proeng.2016.11.635

United Nations Development Programme (UNDP). (2009). *Handbook on planning, monitoring and evaluation for development results.* New York, USA: UNDP.

Waithera, L. & Wanyoike, D. M. (2015). Influence of project monitoring and evaluation on performance of youth funded agribusiness projects in Bahati Sub-County, Nakuru, Kenya. *International Journal of Economics, Commerce and Management,* 3(11), pp. 375–394.

Part III

Communication and leadership in monitoring and evaluation

7 Aspects of communication in monitoring and evaluation

7.1 Abstract

Communication is recognized as critical in the entire life cycle of project implementation. While it is imperative to communicate effectively among the various stakeholders such as the architect, quantity surveyor, engineers, donors and clients during monitoring and evaluation (M&E), communication is less discussed in the M&E literature. What the architect is communicating through his designs must be well and clearly understood and translated to ensure smooth implementation of M&E. Undoubtedly, extant literature has revealed communication as a major setback in the monitoring and evaluation of projects. This chapter argues that effective communication is necessary for effective M&E implementation. Also, communication should be viewed as a multi-way activity involving all project factors in the M&E chain and not the originator and implementers (sender and receiver) and, hence, information must be communicated devoid of ambiguities as well as doubts in expectations. Finally, communication must be targeted to ensure relevant information is delivered timely and to the right audience.

7.2 Introduction

It is universally perceived that all living things, i.e. plants, animals or human beings, communicate through either sound, gestures, speech or movement to express their feelings, problems or understanding of the environment (Goulden, 1992; Jablin & Sias, 2001), hence, the importance of communication in every endeavor of life. In the construction project setting, communication has also been conceptualized variedly (see Dainty, Moore & Murray, 2006; Emmitt and Gorse, 2007; Hoezen et al., 2010; Liu, 2009; Sonnenwald, 1996; Thomas et al. 1998, 1999; Xie et al., 2010). Also, in construction, the M&E process largely requires the sharing of large volumes of information and documents to ensure processes and tasks are accomplished (Cui et al., 2018; Kwofie, Aigbavboa & Thwala, 2019). Characteristic of a typical construction setting where teamwork is essential amongst project stakeholders, communication can be a valuable tool to influence personal, teams and organizational relationships (Pietroforte, 1997; Popple and Towndrow, 1994). In this chapter, a key variable to ensure effective

M&E, i.e. an understanding of the concept of communication in an M&E setting as well as the main attributes of the communication process are theoretically placed in perspective.

7.3 Understanding monitoring and evaluation communication

Communication is necessary in all fields of practice, including the construction industry where M&E of projects takes place throughout the life of the project – even after completion. M&E is a collaborative activity that relies on the effective interaction between all project stakeholders. Project M&E brings together all forms of stakeholders to interact to accomplish and achieve the desired outcome of projects. This makes effective communication non-negotiable in the implementation of M&E as it creates an opportunity for interaction between stakeholders. Project successes as well as challenges need to be communicated for solutions to be attained. Information regarding progress, quality and cost are of great interest to international donors and needs to be communicated. Stakeholders interact to share meaning for greater understanding (Otter & Emmitt, 2008) of the contract and to effectively implement projects at a reasonable cost and time without compromising on quality. Overall, effective communication among project stakeholders is necessary for project success (Lohiya, 2010). Hence, the need to ensure effective communication ensues in the M&E process.

To best understand what communication is, it is imperative to deliberate the meaning of communication. Owing to the multi-dimensional and nebulous nature of the word "communication", it is difficult to have a generalized definition of communication (Dainty et al., 2006). However, a plethora of definitions exists from different authors with different perspectives which attribute different meanings to different people in different situations. First of all, communication is discussed as a process owing to its dynamic nature and the sharing of ideas, goals and opinions (Norouzi, Shabak, Embi & Khan, 2015). It is also described in terms of its function and behavior and, finally, as an interface.

As early as 1953, Hovland, Janis and Kelley (1953) described communication as a process whereby people sent a stimulus purposely to change or to affect the behavior of other people. Metaphorically, Dainty et al. (2006) describe communication as a pipeline through which information is sent to a receiver by a sender. Similarly, communication is the process which explains who says what, through which medium or channel, to whom the information is sent and with what effect (Lasswell, 1948). It is also a process of divulging knowledge from one person to another. In more recent studies, there is evidence of the evolution of the understanding of communication. The definition of communication has become more complex to help address the growing complexity of human activities. Communication has been defined more recently by Perumal and Abu Bakar (2011) as a process of encoding information by a sender and sending the same through a channel to be decoded by the receiver and providing feedback subsequently. It is also the exchange of information or data between a sender and a receiver through writing, speech or signs and the interpretation of the meaning between the parties involved in the process (Adler, 2003; Norouzi et al., 2015; Perumal et al., 2011). Communication is further broken down into content and relationship; content describes the information transmitted whereas the relationship talks about the dynamics between persons involved in the communication (Corcoran, 2007). This separation helps manage the content being transferred in the communication process and also the relationship between the sender and the receiver being well established (Corcoran, 2007).

Secondly, communication has been defined in relation to functions and behaviors (Norouzi et al., 2015). Pietroforte (1997) pointed out that the role of communication is to encourage the accomplishment of shared or joint objectives. It is therefore an avenue to influencing the relationship between parties to achieve shared objectives. Otter and Emmitt (2008) further define communication as human behaviour that facilitates the sharing of meaning and that takes place in a social context. This social relationship is observed as a major characteristic of every communication process where assumptions and expectations of parties can be explored and even explained (Norouzi et al., 2015).

Finally, communication can be explained as an interface. The evolution of communication models over the years has produced sophisticated and computer-mediated communication interfaces such as telephones, emails, fax and video conferencing for communicating amongst parties. In the M&E practice, however, automated cameras, geographical information systems, remote-sensing technologies and global positioning systems have evolved to enhance communication. The use of the right media guarantees effective communication (Ean, 2011)

Communication is therefore perceived as a twin process (two-way process), function or interface for information transfer between people. The major ingredients in every communication system are the presence of a sender of the information, a medium through which information is transmitted, a receiver and a feedback mechanism (Al-Fedaghi, Alsaqa & Fadel, 2009). Encoding and decoding of the information are very critical for sustaining meaning throughout the communication process. Information may include sending verbal or non-verbal messages and receiving feedback as a response. Communication is pervasive in the construction industry and its contribution to effective M&E cannot be overemphasized. Within the M&E context, this could be to convey instructions to influence the behavior and performance of other stakeholders such as the contractor or may involve the exchange of or request for information among clients, consultants and donor agencies.

7.3.1 Types of communication

Communicating amongst project stakeholders comes in several types and forms. Literature acknowledges four main types of communication, namely intrapersonal, interpersonal, mass and group communication. These communication types can take the form of verbal communication to exchange ideas and information at meetings, conferences or through telephone conversations (Perumal & Abu Bakar, 2011). Perumal and Abu Bakar (2011) further describe the other form of communication as non-verbal, thus written communication through handwritten or computer output, reports, pictures, emails and minutes. Communication can also take the form of internal or external communication (Perumal & Abu Bakar, 2011).

Communication does not just happen; there are established communication channels for different fields. In contract administration, information flows to some stakeholders are limited or may not even have a formal communication channel. The M&E information is also released based on the level of involvement of the stakeholder. Donor agencies who have an interest in the overall project success are predisposed to cost data of the project, just as beneficiary stakeholders are interested in the completion of the project on time. For monitoring to achieve its aim of participation and accountability, information such as project deliverables, client expectation and contract summary should well be communicated to all teams and stakeholders (Lohiya, 2010). Also, barriers to information flow must be eliminated even though particular interests may exist. Thus, all stakeholders must have an equal opportunity to any information they require on the project.

7.3.1.1 Intrapersonal communication

Intrapersonal communication describes the type of communication within oneself or each stakeholder. In project M&E, however, this form of communication can be described as communication within each identified stakeholder. This is seen in the communication among the quantity-surveying team or the architectural department or generally within the project consultants as a unique stakeholder in the project. It is essential for categories of stakeholders such as the quantity-surveying team to communicate internally among themselves in dealing with cost-related matters in the M&E process. The same can be said for the architectural and the engineering units.

7.3.1.2 Interpersonal communication

Interpersonal communication refers to information transfer between two people. Dainty et al. (2006) posit that this type of communication is typically subjective in nature and value-laden. Though all stakeholders are involved in project implementation, Tengan and Aigbavboa (2017) argue that their involvement in M&E is limited to the three core stakeholders who have interest and influence in the project; thus, consultant, contractor and client. Interpersonal communication provides a face-to-face interaction between stakeholders and enables understanding, stimulates the sharing of expert knowledge and encourages team building (Otter & Emmitt, 2008). Project stakeholders communicate with each other by providing project information and receiving the same from them. Otter and Emmitt (2008) further indicate that interpersonal communication affords stakeholders the opportunity for understanding since multiple meanings could be derived from the non-verbal gestures such as body language to support the verbal acts. Established communication between the project manager and the client is an example of an interpersonal communication. This is also seen in instructions issued to contractors or the progress report submitted to the client by the project manager. This type of communication is the most frequent in the M&E of construction projects.

7.3.1.3 Mass communication

Mass communication is described as an interaction between a large, heterogeneous, assorted and anonymous audience. The number of the audience involved in this communication is not known and involves types of people, there is no limitation in geographical location and communicators are not known in person at all. Therefore, there is the need for a communication channel that produces and transmits information to such large audience. Mass communication takes the form of three major categories based on its physical form; print media (magazines, periodicals and books), electronic (radio and television) and digital media like Internet facilities. In the day-to-day implementation of the project (M&E) beyond communication within specific stakeholders or among stakeholders, some information is very relevant to the public and visitors to the project site or location. Communicating safety measures, for instance, is crucial for passers-by who may not be known or expected. If project activities affect the general public such as in the case of the maintenance of a major bridge, it would be imperative to communicate alternative routes to the public at large to avoid accidents occurring.

7.3.1.4 Group communication

Group communication, also referred to as team communication, refers to communication that ensues between three or more people or project stakeholders with a common goal or objective. This communication occurs in a unique or defined group. This takes the form of face-to-face communication or with the aid of computer applications such as Skype. It provides an opportunity for all stakeholders, namely the contractor, sub-contractor, client, consultant, local service authorities, implementation organization and beneficiary community representation to receive or provide necessary information on the state of project implementation and to collectively make decisions. In a project environment, this type of communication is envisaged during site meetings where the work of all stakeholders, particularly that of the contractor and sub-contractor (nominated), is closely scrutinized to ensure adherence to project specifications.

7.3.2 Communication models

A communication model is a graphic representation of an ideal and orderly process of communication and serves as a standard tool for communication (Al-Fedaghi et al., 2009). These communication models help us shape and understand the way stakeholders or team members communicate (Dainty et al., 2006), while the complex nature of communication is simplified. An appreciation of communication models provides alternative approaches (discoveries) to effective communication

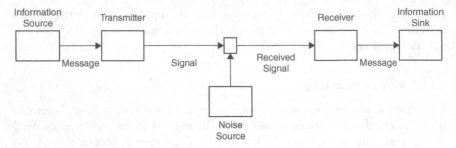

Figure 7.1 Shannon and Weaver's model of communication.

Source: Shannon, 1948

in complex environments such as the construction industry. A variety of communication models has been professed in several studies. Older models, however, have been improved upon to ensure that complex situations in the global world and fields are circumvented. It is argued that the most widely adopted communication model in many fields is Shannon's model of communication which has influenced some established communication models (Al-Fedaghi et al., 2009). Identified models of communication are discussed below.

7.3.2.1 Shannon and Weaver's model of communication, 1948

In 1948, Claude Shannon originally developed a communication model to understand how information bits are transmitted from a source to the destination in a telephone conversation. Shannon's (1948) model of communication is regarded as the most influential among earlier communication models and presented in Figure 7.1. Subsequently, Warren Weaver (Shannon's colleague) complemented Shannon's model with the introduction of a mechanism in the receiver to correct the differences between the transmitted and received signal, thus the feedback element. The main elements outlined in this model are the information source, a transmitter (encoder), channel, noise, receiver (decoder), destination and feedback. Shannon and Weaver's model of communication attempted to reduce the communication process to a set of mathematical formulas with the aim of achieving 100% efficiency in the communication process. Subsequent to the model, other concepts were introduced to upscale the existing model which include the measure of uncertainty in the system (entropy), redundancy which describes the degree of repetition and its effect on the understanding of the transmitted information, the measure of irrelevant information to the message (noise) and the capacity of the channel to receive and transmit the information (maximum capacity).

7.3.2.2 Lasswell's model of communication, 1948

Figure 7.2 illustrates Lasswell's model of communication. Lasswell's (1948) model of communication focused on mass communication. He therefore developed a

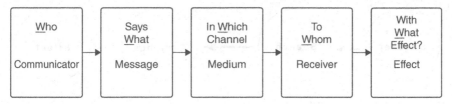

Figure 7.2 Lasswell's model of communication.

Source: Lasswell (1948)

verbal model to describe the communication process. This model explained five elements making up the communication process. The model stressed Who – says What – in Which channel – to Whom and with What effect. A major contribution of Lasswell's communication model is the concept of effect. It is also an easy communication process and it suits almost all types of communication. However, feedback mechanisms were not addressed in his model, making the effect of the communication difficult to measure. It also falls flat in addressing communication barriers such as noise. It is a linear or one-way communication model with multiple mediums of carrying the same information.

7.3.2.3 Osgood-Schramm's model of communication

Schramm Wilbur is admired as among the founding fathers of mass communication. This model presented in Figure 7.3 illustrates the importance of interpersonal communication. The model describes the circular nature of communication which suggest that at every point in the communication process, the sender takes turns to be the receiver and vice versa. Osgood–Schramm's model is made up of six elements of communication, namely source, sender, encoder, message, channel, decoder and receiver. This model integrates an essential element of interpretation. That means the receiver does not only decode the message but also interprets the message and makes meaning out of it. Emphasis is also placed

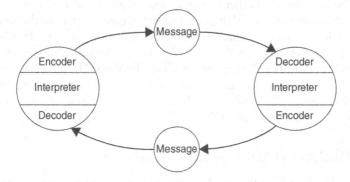

Figure 7.3 Osgood–Schramm's model of communication.

Source: Schramm (1954)

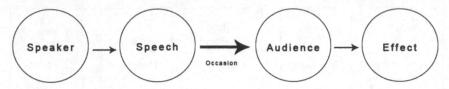

Figure 7.4 Aristotle's model of communication.

Source: Fritz Heider in 1958

on the message sent and message received rather than the channel of message transmission. Osgood–Schramm describe their model as endless since information does not end with the receiver; it continuously moves to the sender in the form of feedback and this process ensures that communication is refined to eliminate semantic noise and promotes knowledge expansion among the stakeholders while maintaining the flow of communication. The Osgood–Schramm's model of communication encourages education and learning among senders and receivers in an environment where either sender or receiver has little knowledge about the message or information being sent or discussed.

7.3.2.4 Aristotle's model of communication

The Greek philosopher, Aristotle, writing 300 years before the birth of Christ, developed a linear model of communication based on rhetorical perspective (see Figure 7.4). This involved communicating to the masses (audience) with the aim of influencing and persuading them. His model was developed around three elements, namely Speaker-Speech-Audience. However, some other studies refer to a five-element model which introduces two more elements comprising Speaker-Speech-Occasion-Audience-Effect. This model is accepted as the first communication model.

In terms of a speaker-centered model, Aristotle found the role of audience and listeners imperative and, as such, described them as the last mile of communication having the answers to whether or not communication had taken place. Dainty et al. (2006), however, argue that until feedback is given back from the receiver to the sender, it remains uncertain whether communication has indeed taken place. The model further recognizes the *occasion* for the communication. Aristotle's model of communication did not come without shortfalls. The model was criticized for its lack of a feedback mechanism (making it a one-way communication process, namely from speaker to audience) and the non-consideration of communication failures such as noise and barriers. It was also limited to public speaking.

7.3.2.5 David Berlo's S-M-C-R model of communication

David Berlo's S-M-C-R model of communication is presented in Figure 7.5. The model is centered on sender, message, channel and receiver. In Berlo's view, communication is about sending or transmitting the message and not

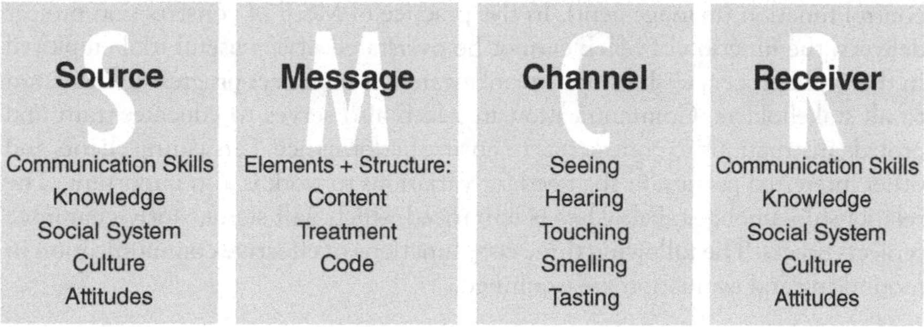

Source	Message	Channel	Receiver
Communication Skills	Elements + Structure:	Seeing	Communication Skills
Knowledge	Content	Hearing	Knowledge
Social System	Treatment	Touching	Social System
Culture	Code	Smelling	Culture
Attitudes		Tasting	Attitudes

Figure 7.5 Berlo's S-M-C-R model of communication.

Source: David Berlo's S-M-C-R Model (1960)

the transmission of meanings. He further acknowledges that meaning is not in the message that is sent but rather it is personal to the sender and receiver; thus, the meaning is adduced at the encoding and decoding stages of the communication process. Persons or stakeholders may have a similar understanding, yet will make different meaning of the information unless they have similar encounters. Besides, meanings are dynamic. For instance, meanings change as one's experience changes. The pre-encoded message sent could contrast to a lesser or more noteworthy degree from the received decoded message. He argues that the importance that is given by sender of a message might be altogether different from the significance the receiver appends to the same message because of different encounters, skillsets, attitudes, knowledge social systems and cultures.

A well-structured message is defined by its content, elements, treatment, structure and code as illustrated in Figure 7.5. The channel in Berlo's model includes the sense of hearing, touching, tasting, smelling and seeing through which signals or messages can be transmitted to the receiver. Berlo's model of communication, however, did not account for noise or barriers in the communication process. Noise is classified as physical or semantics. Physical noise could be a distraction from other activities ongoing at the site. Semantic noise is that which emanates from the message sender or both. The language or codes used in the communication may be too technical or jargonistic which leads to loss of attention or misunderstanding. Berlo's model fell short of this challenge and pre-supposes that effective communication will take place as if both sender and receiver are both on the same level of knowledge. Finally, the element of feedback is not considered in his model.

7.4 Functions of communication in monitoring and evaluation

Communication performs several functions in social life and the environment. Such functions may include educating, informing, cultural promotion, social contact, integration, stimulation, counselling and expression of emotions and as a control function (management). In the practice of M&E of construction project delivery, the function of M&E cannot be overstated. It is a useful tool employed in the M&E process to share and create awareness of project progress information to all stakeholders. Communication in M&E also serves to educate, train and provide instructions to contractors to ensure performance. Persuading clients and other interested parties on the need for variations to work is also important. The relationship among stakeholders is enhanced which will significantly guarantee project success. The following three core functions of effective communication in monitoring and evaluation are examined.

7.4.1 *Information sharing and awareness creation*

According to Otieno (2000), effective and timely communication of information to users is a major strength of management in project implementation. It has also been established in the literature that a major challenge in M&E is the utilization of M&E findings (Tengan & Aigbavboa, 2016). Indeed, Otieno (2000) describes communication in project management as the catalyst in achieving the desired project objectives. M&E data and findings are relevant to all stakeholders and therefore the need for effective tools to disseminate such information to all interested stakeholders. Interested parties include direct and indirect stakeholders, the external funding organization and the general citizenry whose taxes are being used for developmental projects and who must rightly be informed about the progress (accountability). This sustains stakeholder confidence and ensures their constant desire to support future development (Umhlaba Development Services, 2017). In sharing information to create awareness of project progress at every stage, care must be taken to ensure such findings are accurate, timely and complete (Murray, Tookey, Langford & Hardcastle, 2000). This may take many forms such as verbal communication at a formal site or management/stakeholder meeting, submission of reports through emails and courier services. To ensure effective communication of M&E findings and information, it is appropriate to agree on the frequency of communication, how and what form the communication should take and the timelines to expect feedback for decision-making purposes.

7.4.2 *Education and training*

Organization learning is made possible and enhanced when effective communication is ensured in the M&E process. Projects are unique in several ways, including scope and implementation. Effective M&E will reveal challenges in project implementation which will help in making decisions and in adopting strategies in overcoming such challenges in future projects which may have similar features. M&E provides an opportunity to educate the public and labor involved in the project. Safety information is communicated widely to reduce fatalities at and around the site. M&E findings can also call for the training of team members or some stakeholders on some practices relevant to the project implementation.

7.4.3 Persuasion

Project implementation brings together several stakeholders to undertake such important activity like monitoring and evaluation. The M&E activity is usually led by the consulting team or the project manager. This leading role comes with numerous challenges. These stakeholders do not all report to the project manager; it is, therefore, essential to effectively communicate monitoring information to stakeholders at all levels particularly with the aim to influence them. Stakeholders get motivated by what is effectively communicated to them which will strengthen stakeholder relationship, commitment and participation in the project. This if done efficiently will ensure that project stakeholders are on the same level of understanding of what is happening on the project. The effect is assurance of project success ultimately.

7.5 The use of information communication technology (ICT) in construction project monitoring and evaluation

The effectiveness of communication on projects becomes increasingly challenged when the project scope increases when multiple concurrent activities need to be accomplished. It becomes imperative therefore to adopt strategies and technologies to mitigate any challenges to communication arising from increased scope and multiplicity of tasks on the project. It is important to note that while the benefits of using information and communication in construction have been appreciated and widely known, Peansupap and Walker (2004) report on the industry's slow attitude towards adoption, confirming the poor adoption to change of many construction industries. Also, there is a culture of oral and face-to-face communication in the construction industry (Ballan & El-Diraby, 2011). Additionally, innovative technology adoption is said to contribute to productivity, quality and safety improvement in construction (Sepasgozar & Bernold, 2013).

In the M&E of projects for success, any delay in communicating information can have dire consequences on the performance of projects. For example, communication regarding the clarification of instructions given by project manager, if not received and acted upon timeously, can be the basis for an extension of time or grounds for demolition and rework (increased project cost). Sunjka and Jacob (2013) attest to the misunderstanding and misrepresentation of facts on projects in the Niger Delta, Nigeria due to the poor or inadequate communication between parties. On the back of the challenges in technology adoption in the construction industry and the lack thereof in monitoring and evaluation of projects, Ballan et al. (2011) make some recommendations which include ensuring that the content of the communication is meaningful to the construction industry, that there is ready access at all times to project information and, finally, ensuring that communication systems are user-friendly.

The utilization of technology such as information and communication technology (ICT) in M&E is underdeveloped even though the potential impact on the entire project cycle, project diagnosis, planning, implementation, monitoring, evaluation, reporting, sharing and learning is enormous (Raftree & Bamberger, 2014). According to Bohn and Teizer (2009), adoption and implementation of technology in construction have largely been unsuccessful as a result of the different understanding of the benefits and limitations of technology use in construction. M&E of the project is continuously undertaken throughout the project implementation process, making the practice time-consuming as well as a manual process with visual inspections adopted. The M&E activities such as tracking and updating project schedule can be assisted with technology such as the use of high-resolution automated cameras that can provide construction management and other users with project site images and videos (Bohn & Teizer, 2009). In using cameras, Senior and Swanberg-Mee (1997) argue that several imbalances such as time wasting, task completion time and inefficiencies can be recognized and adjusted for resource optimization. This is made possible through tracking of the workforce, equipment and materials' inventory (Senior & Swanberg-Mee, 1997).

Technologies may operate either on wireless signals or optical measurement. Wireless signal technologies include the global positioning system (GPS) for machine site utilization and position control (Navon, 2008), the radio frequency identification which helps in locating and tracking material on- and off-site (Jaselskis & Gao, 2003) and the ultra-wideband for real-time resource tracking and work zone safety (Teizer et al., 2005). The laser rangefinders for machine guidance and position and laser scanners for three-dimensional point cloud measurement are regarded as optimal measurement technologies (Bosche & Haas, 2008; Lytle & Saidi, 2007). In other studies, mobile technology (mobile phones) have been adopted for M&E of international health and development programmes in developing countries to fast-track data collection (Bruce, 2013). Technology adoption in M&E has proven to contribute to the successful implementation of projects by eliminating rework and increased measurement accuracy (Bohn & Teizer, 2009). Further, data collection and handling are made easy and straightforward.

7.6 Benefits of monitoring and evaluation communication

Several benefits are observed in literature from the diverse field as a result of effective communication. There is a benefit to the individuals, the process and the organization. Likewise, in the construction industry, effective communication has enormous benefits in the M&E of the project and so cannot be overstated. Effective communication contributes to the achievement of coordinated results. Coordinated results in effective M&E are seen in the ultimate achievement of project success. This can only be possible if there is an effective communication among stakeholders. Managing change in construction project delivery has been a daunting task for project stakeholders. Through effective communication during the M&E of construction projects, organizational change is guaranteed. Husain (2013) established the positive contribution of effective communication to organizational change. He further argues that for effective communication to bring about organizational change, participation and commitment of employees, trust and feedback mechanisms in the communication process are critical and must be managed well.

Effective communication in the M&E process reduces the frequency of misunderstandings and consequent errors on the part of contractors. Thus, if architectural and engineering design details, bills of quantities and other forms of communication are clear and unambiguous, errors and re-work are eliminated to the minimum with significant consequences on project cost and schedule. Dewulf, Hoezen and Reymen (2006) posit that the result of effective communication is improvement in communication relationship between stakeholders which could reduce project failures. He further argues that a more open communication at all levels of the project cycle could also lead to innovation and better technical solutions to challenges that may come up. Again, effective communication at the early and briefing stages of the project could positively influence the quality of the project as desired by all stakeholders and lead to better decision making (Hoezen et al., 2006).

7.7 Barriers to monitoring and evaluation communication

The nature and complexity of the construction industry pose challenges to every activity aimed at achieving successful project delivery. Likewise, in the implementation of M&E, communication barriers do exist. Identifying these barriers will ensure adequate measures are developed in the communication process to guarantee effective communication during M&E. Key barriers to effective communication are discussed as follows. Hoezen et al. (2006) argue that stakeholder interest in project delivery influences how they communicate amongst themselves, stressing that opposing interests could lead to hidden agendas which will lead to restricted communication. Cross-cultural barriers are significant barriers to effective communication (Adler, 2003). Owing to the diversity of stakeholders engaged in the M&E of projects as well as their occupational cultures and professional background such as quantity-surveying, architecture, engineering and accounting, unskilled labor and craftsmen/women communicating in a manner that will be understood by all other professionals as well as skilled and unskilled labour to ensure effective communication is important.

The transient and dynamic nature of the construction industry's activities necessitates the moving of the workforce from one project site to another in different geographical locations. As described by Dainty et al. (2006), this could result in a cultural barrier. Communicating project information also to foreign donor-funding organizations ought to be clear in order not to miscommunicate owing to cultural differences. Jargon and semantics have been identified as critical barriers to effective communication (Dainty et al., 2006). The technical nature of the construction industry has led to the adoption of formal and informal languages (common language) among professions for easy communication (Delisle & Olson, 2004). These terminologies used by project managers or specific stakeholder such as the quantity surveyor could be misleading and not understood by other stakeholders, thereby obstructing communication. Dainty et al. (2006) argue that documents such as drawings, specifications, method statements and some other project documents could be the source of miscommunication. Noise is also a major barrier to verbal communication (Dainty et al., 2006). As a result of the construction industry's activities and processes on site, effective communication between supervisors and site engineers and laborers on the project can be affected.

Poor communication practices during M&E and failing to overcome the barriers outlined above can have dire consequences for construction project success. It is therefore imperative to work towards addressing the communication challenges in the M&E of construction project delivery.

7.8 Achieving effective monitoring and evaluation communication

Communication has been understood to be a cumbersome process and if not handled well, it will render all efforts in the M&E of projects ineffective. Indeed, the lack of effective communication has been suggested by several studies as a major challenge in the M&E of projects (Mugambi & Kanda, 2013). Hence, content or information being transmitted and the relationship between stakeholders are critical and must be managed well. Windapo, Odediran & Akintona (2015) argue that to achieve project success, the relationship between the project manager and other stakeholders is significant and must be enhanced through effective communication.

Mugambi and Kanda (2013) stressed that for communication to be effective and to significantly influence the M&E of projects, the communicator, that is, the project manager or leader in most cases, must be knowledgeable and must possess excellent interpersonal and communication skills. In that case, information to be transmitted will be sufficiently encoded before engaging any communication for proper understanding by all other stakeholders. Communication is a two-way process where information emanating from the project manager is received by all other stakeholders; hence the need for good listening skills by the recipients of the information. This, however, must be done through appropriate channels of communication.

Communication is effective when received in its right form and context and acted upon accordingly and feedback is communicated at the right time. Receipt of information in its right form and context is achieved through proper communication channels. In project M&E, the most common channel of communication is the form that takes the face-to-face approach where one visits the project site and gives verbal instructions or deliver it in a written form in the site instruction book. The face-to-face form can also be seen during site meetings where all stakeholders are present and project information is shared with appropriate feedback given. At site meetings, communication in the form of progress report and works programme (computer-aided output) is received from contractors. In some situations, before the site meeting, this communication (progress report and works programme) is sent through the Internet to the appropriate stakeholders by the project manager. Meeting invitations and previous meetings' minutes emanating from the project manager are also transmitted through the Internet with the help of computers.

Norouzi et al. (2015) posit that the effectiveness and the right use of communication media contribute to the delivery of information at the right time, form and context, hence effective communication. Access to project information by stakeholders and parties is important in promoting effective communication. However, care must be taken to ensure the right information is given to the right persons or stakeholders to avoid all ambiguity and distortions. The timeliness, accuracy and completeness of the release of information are critical (Murray et al., 2000). Stage information indicates whether the project is on track and needs to proceed or necessary revisions are required before moving on to the next stage of the project. Such stage information ensures that projects are delivered on schedule (Njama, 2015). Murray et al. (2000) in a study to identify project communication variables through a case of the USA and the UK construction industries' perceptions stressed the level of understanding of information communicated as well as barriers to and procedures for the communication of project information. The overarching communication barriers must be a critical factor impeding effective communication, hence the need to eliminate or reduce these to the barest minimum. Existing formally defined procedures outlining the scope and method of communication will serve as a guide for effective communication. Questions about how communication has been done in the past and what challenges existed will ensure the right procedures and methods are adopted for effective project-based communication.

Perumal and Abu Bakar (2011) argue strongly for the need to standardize documents for communication. This in their view will encourage effective communication among stakeholders in the construction industry. Proper communication structures will also serve as a panacea to communication lapses in M&E (Perumal & Abu Bakar, 2011). There is a need to establish appropriate levels of communication. However, care must be taken to ensure that these communication structures do not create barriers to information flow. Communication is understood within the environment or setting in which it occurs (Otter & Emmitt, 2008). The implication is that official communication should be communicated within the appropriate environment to receive the right attention. Finally, the reporting system adopted for the M&E process will significantly influence the communication process. Projects fail for many reasons and one significant reason is ineffective project reporting systems (KPMG, 2014).

Communication in the construction industry is largely traditional where it is human or people-centred, that is, the face-to-face approach. Adoption of technology in the M&E of the construction projects has become topical. BIM, for example, has been described by Goh, Goh, Toh and Peniel Ang (2014) as an M&E tool that provides a solution to the communication problems in the management of projects. Other advanced technologies such as the geographical information systems (GIS) and computer-mediated communication significantly improve communication among parties (Poku & Arditi, 2006) and the management of project information. Communication is sent swiftly at the press of a button and delivered to one or many recipients, making the delivery of the communication and feedback mechanisms fast and efficient. It is therefore necessary to adopt technologies such as BIM to enhance communication to deliver the desired results of every M&E.

Summary

The chapter discussed generally the need for proper communication in construction project management, particularly in the M&E process. The types and mode (model) of communication are discussed. Also the purpose of communication in M&E is explained to include information sharing and awareness creation, education and training and persuasion. Furthermore, the benefits and barriers of M&E implementation are reviewed. Some approaches to achieving effective communication in the M&E process are discussed as well. The next chapter introduces yet another variable essential for effective implementation of M&E relating to construction which is monitoring and evaluation leadership.

References

Adler, N. J. (2003). Communicating across cultural barriers. *Interkulturelle Kommunikation: Texteund Übungen zum interkulturellen Handeln*. Sternenfels: Wissenschaft & Praxis, pp. 247–276.
Al-Fedaghi, S., Alsaqa, A. & Fadel, Z. (2009). Conceptual model for communication. *International Journal of Computer Science and Information Security*, 6(2), pp. 29–41.
Ballan, S. & El-Diraby, T. E. (2011). A value map for communication systems in construction. *Journal of Information Technology in Construction (ITcon)*, 16, pp. 745–760.
Bohn, J. S. & Teizer, J. (2009). Benefits and barriers of construction project monitoring using high-resolution automated cameras. *Journal of Construction Engineering and Management*, 136(6), pp. 632–640.
Bosche, F. N. & Haas, C. T. (2008). Automated retrieval of project three-dimensional CAD objects in range point clouds to support automated dimensional QA/QC. *ITcon*, 13, p. 16.
Bruce, K. (2013). *Use of mobile technology for monitoring and evaluation in international health and development programs*. ProQuest LLC. Available online at: http://www.proquest.com/en-US/products/dissertations/individuals.shtml.
Corcoran, N. (2007). Theories and models in communicating health messages. *Communicating health: Strategies for health promotion*, pp. 5–31.
Cui, C., Liu, Y., Hope, A. & Wang, J. (2018). Review of studies on the public–private partnerships (PPP) for infrastructure projects. *International Journal of Project Management*, 36(5), pp. 773–794.
Dainty, A., Moore, D. & Murray, M. (2006). *Communication in construction-theory and practice*. Milton Park, Abingdon, Oxon: Taylor & Francis.

Delisle, C. L. & Olson, D. (2004). Would the real project management language please stand up? *International Journal of Project Management*, 22(4), pp. 327–337, doi:10.1016/S0263-7863(03)00072-3

Ean, L. C. (2011). Computer-mediated communication and organisational communication: The use of new communication technology in the workplace. *The Journal of the South-East Asia Research Centre for Communication and Humanities*, 3, pp. 1–12.

Emmitt, S. & Gorse, C. A. (2007). *Communication in construction teams*. London: Taylor & Francis.

Goh, K., Goh, H., Toh, S. & Peniel Ang, S. (2014). Enhancing communication in construction industry through BIM. *Presented at the 11th International Conference on Innovation & Management*.

Goulden, N. (1992). Theory and vocabulary for communication assessments. *Communication Education*, 41(3), pp. 258–269.

Hoezen, M. E. L., Reymen, I. & Dewulf, G. P. (2006). The problem of communication in construction The Netherlands: University of Twente.

Hoezen, M., Rutten, J. V., Voordijk, H. & Dewulf, G. (2010). Towards better customized service-led contracts through the competitive dialogue procedure. *Construction Management and Economics*, 28(11), pp. 1177–1186.

Hovland, C., Janis, I. & Kelley, H. H. (1953). *Communication and persuasion: Psychological studies of opinion change*. New Haven, CT: Yale University Press.

Husain, Z. (2013). Effective communication brings successful organizational change. *The Business & Management Review*, 3(2), p. 43.

Jablin, F. & Sias, P. (2001). Communication competence. In: Jablin, F. and Putnam, L. (Eds.). *The new handbook of organizational communication*. Thousand Oaks, CA: SAGE.

Jaselskis, E. J. & Gao, Z. (2003). Pilot study on laser-scanning technology for transportation projects. *Presented at the Mid-Continent Transportation Research Symposium*. Ames, Iowa: Iowa State University.

Klynveld Peat Marwick Goerdele (KPMG) (2014). *Effective reporting for construction projects: increasing the likelihood of project success*. Project Advisory.

Kwofie, T. E., Aigbavboa, C. O. & Thwala, W. D. (2019). Communication performance challenges in PPP projects: Cases of Ghana and South Africa. *Built Environment Project and Asset Management*, 9(5), pp. 628–641.

Lasswell, H. D. (1948). *The structure and function of communication in society. In L. Bryson(Ed.), The communication of ideas: A series of addresses* (pp. 37–51). New York, NY: Institute for Religious and Social Studies.

Liu, Y. (2009). *Critical factors for managing project team communication at the construction stage*. Ph.D. Thesis. Submitted to the Polytechnic University of Hong Kong.

Lohiya, G. (2010). *Team building in project management practice in the UAE construction industry*.

Lytle, A. M. & Saidi, K. S. (2007). NIST research in autonomous construction. *Autonomous Robots*, 22(3), pp. 211–221, doi:10.1007/s10514-006-9003-x

Mugambi, F. & Kanda, E. (2013). Determinants of effective monitoring and evaluation of strategy implementation of community-based projects. *International Journal of Innovative Research and Development*, 2(11).

Murray, M., Tookey, J., Langford, D. & Hardcastle, C. (2000). Project communication variables: A comparative study of US and UK industry perceptions.

Navon, R. (2008). Automated productivity control of labor and road construction. *Presented at the 25th International Symposium on Automation and Robotics in Construction*. Vilnius, Lithuania: Vilnius Gediminas Technical University Publishing House Technika, pp. 29–32.

Njama, A. W. (2015). *Determinants of effectiveness of a monitoring and evaluation system for projects: A case of Amref Kenya WASH programme.* Kenya: University of Nairobi.

Norouzi, N., Shabak, M., Embi, M. R. B. & Khan, T. H. (2015). The architect, the client and effective communication in architectural design practice. *Procedia – Social and Behavioral Sciences,* 172, pp. 635–642, doi:10.1016/j.sbspro.2015.01.413

Otieno, F. A. O. (2000). The roles of monitoring and evaluation in projects. In: *2nd International Conference on Construction in Developing Countries: Challenges Facing the Construction Industry in Developing Countries.* pp. 15–17.

Otter, A. & Emmitt, S. (2008). Design team communication and design task complexity: The preference for dialogues. *Architectural Engineering and Design Management,* 4(2), pp. 121–129.

Peansupap, V. & Walker, D. H. (2004). Strategic adoption of information and communication technology (ICT): Case studies of construction contractors. In: Khosrowshahi, F. (Ed.). *20th Annual Association of Researchers in Construction Management (ARCOM) Conference.* Edinburgh, Scotland: Heriot-Watt University, pp. 1235–1245.

Perumal, V. R. & Abu Bakar, A. H. (2011). The needs for standardization of documents towards an efficient communication in the construction industry. *World Applied Science Journal,* 12(9), pp. 1988–1995.

Pietroforte, R. (1997). Communication and governance in the building process. *Construction Management and Economics,* 15(1), pp. 71–82, doi:10.1080/014461997373123

Poku, S. E. & Arditi, D. (2006). Construction scheduling and progress control using geographical information systems. *Journal of Computing in Civil Engineering,* 20(5), pp. 351–360.

Popple, G. W. & Towndrow, S. P. (1994). Communication for the construction industry. *BT Technology Journal,* 13(3), pp. 45–50.

Raftree, L. & Bamberger, M. (2014). *Emerging opportunities: Monitoring and evaluation in a tech-enabled world.*

Schramm, W. (1954). How communication works. In: W. Scramm, (Ed.), *The Process and Effects of Mass Communication* (pp. 3–26). Urbana, IL: University of Illinois Press.

Senior, B. A. & Swanberg-Mee, A. (1997). Activity analysis using computer-processed time lapse video: Proceedings of the congress, Minneapolis, Minnesota, October 4–8, 1997. In: *Managing Engineered Construction in Expanding Global Markets.* Presented at the Construction Congress, Reston, Va: ASCE, pp. 462–469.

Sepasgozar, S. M. & Bernold, L. E. (2013). Factors influencing construction technology adoption. In: *19th CIB World Building Congress,* Brisbane.

Shannon, C.E. (1948). A mathematical theory of communication. *Reprinted with corrections from The Bell System Technical Journal,* 27, 379–423, pp. 623–656.

Sonnenwald, D. H. (1996). Communication roles that support collaboration during the design process. *Design Studies,* 17(3), pp. 277–299.

Sunjka, B. P. & Jacob, U. (2013). Significant causes and effects of project delays in the Niger Delta Region, Nigeria. In: *SAIIE25 Proceedings.* Stellenbosch, South Africa: SAIIE, pp. 641–655.

Teizer, J., Kim, C., Haas, C., Liapi, K. & Caldas, C. (2005). Framework for real-time three-dimensional modeling of infrastructure. *Transportation Research Record: Journal of the Transportation Research Board,* 1913, pp. 177–186, doi:10.3141/1913-17

Tengan, C. & Aigbavboa, C. (2016). Evaluating barriers to effective implementation of project monitoring and evaluation in the Ghanaian construction industry. *Procedia Engineering,* 164, pp. 389–394, doi: 10.1016/j.proeng.2016.11.635

Tengan, C. & Aigbavboa, C. (2017). Level of stakeholder engagement and participation in monitoring and evaluation of construction projects in Ghana. *Procedia Engineering*, 196, pp. 630–637, doi: 10.1016/j.proeng.2017.08.051

Thomas, S. R., Tucker, R. & Kelly, W. (1998). Critical communications variables. *Journal of Construction Engineering Management*, 129(1), pp. 58–66.

Thomas, S. R., Tucker, R. & Kelly, W. (1999). Compass: An assessment tool for improving project team performance. *Project Management Journal*, 30(4), pp. 15–23.

Umhlaba Development Services (2017). Introduction to monitoring and evaluation using the logical framework approach. Noswal Hall, Braamfontein, Johannesburg, South Africa

Windapo, A., Odediran, S. & Akintona, R. (2015). Establishing the relationship between construction project managers' skills and project performance. In: *ASC Proceedings of the 51st Annual Conference.*

Xie, C.,Wu, D., Luo, J. & Hu, X. (2010). A case study of multi-team communications in construction design under supply chain partnering. *Supply Chain Management: An International Journal*, 15(5), pp. 363–370.

8 Perspectives in monitoring and evaluation leadership

8.1 Abstract

In a monitoring and evaluation (M&E) environment dominated by co-equals, leadership is thought to offer significant guidance to the practice or operations of an M&E team. Many leadership theories have evolved in response to the effective dealing with the dynamic and complex nature of construction activities and processes. However, there is not much understanding of the attributes, dynamics and context of leadership in the M&E processes. Given that M&E is achieving increasing popularity in the global construction industry, it is thus vital for deeper theoretical and conceptual understanding into leadership attributes, dynamics and context in M&E. The focus of this chapter is on the evolution and leadership styles, the barriers experienced and the means of achieving effective M&E leadership. These insights can serve the basis on which the benefits of effective leadership in M&E for construction project delivery can be enhanced. Leadership competencies such as the ability to inspire others to deliver a task, the collaborative orientation of leaders, exhibiting intelligence and learning agility, practicing accountability and exerting real judgments cannot be overlooked if leadership effectiveness is to be achieved. This is a sure way to improving performance in construction project delivery.

8.2 Introduction

In the M&E of construction projects, different stakeholders, such as consultants (quantity surveyors, architects, engineers), clients, (sub-)contractors and donors with dissimilar interests are brought together to interact with the ultimate objective of achieving project performance indicators within time constraints (Zakaria, Mohamed, Ahzahar & Hashim, 2015). M&E leadership is imperative in the M&E process and is provided at two levels; leadership amongst other key stakeholders and leadership towards labor or artisans. Given this, therefore, leadership is an essential characteristic in the M&E process to ensure improved productivity (Zakaria et al., 2015). Leadership in the M&E of construction projects is traditionally vested in the project manager who may be an individual or a consulting team. He is knowledgeable in construction management and has several

Table 8.1 Catalogue of leadership definitions

Source	Leadership definition
Fiedler (1967: 36)	"Leadership is described as directing and coordinating the work of a group."
Burns (1978: 425)	"Leadership is the reciprocal process of mobilizing persons with certain motives and values, various economic, political and other resources, in the context of competition and conflict, to realize goals independently or mutually held by both leaders and followers."
Bennis (1989: 65)	"The capacity to create a compelling vision and translate it into action and sustain it."
Yukl (1989: 253)	"Leadership involves influencing task objectives and strategies, influencing commitment and compliance in task behavior to achieve these objectives, influencing group maintenance and identification and influencing the culture of an organization."
Bass, Bernard & Avolio (1990)	"The principal dynamic force that motivates and coordinates the organization in the accomplishment of its objectives."
Gardner (1990: 1)	"Leadership is the process of persuasion or example by which an individual (or leadership team) induces a group to pursue objectives held by the leader and his or her followers."
Chemers (1997: 1)	"Leadership is a process of social influence in which one person can enlist the aid and support of others in the accomplishment of a common task."
Vroom et al. (2007: 18)	"...a process of motivating people to work together collaboratively to accomplish great things."

Source: Ofori & Toor, 2012

other inherent characteristics and skills. Leadership arguably is as an important factor in construction project progress monitoring (Zakaria et al., 2015) as it is with construction project management (Ofori & Toor, 2012; Panthi, Farooqui & Ahmed, 2010). The term leadership is omnipresent (Vroom & Jago, 2007) in business and organizational management. However, there is no consensus on a single acceptable definition (Ofori & Toor, 2012). More conveniently, it has been described in many fields by several scholars. A catalogue of definition on leadership is presented in Table 8.1:

The concept of leadership has evolved over several years while generating enormous debates and in most cases, creating confusion and misunderstanding. In the more recent past, leadership has been defined as a process of inducing the activities of individuals and groups to accomplishing common goals (Koech & Namusonge, 2012; Sharma & Jain, 2013). Guided by the above definitions, leadership in monitoring and evaluation can therefore be described as a process whereby project managers can influence other stakeholders and labor by directing, coordinating, mobilizing, motivating as well as persuading them in a project environment to achieve project success. However, the leadership in bringing about project success in the M&E process is not limited to the project manager but also the commitment of all other stakeholders engaged on the project as well as the leadership situation.

8.3 Types of leadership styles

Many types of leadership styles have been observed in literature with their impact being largely reported. Bhatti, Maitlo, Shaikh, Hashmi, and Shaikh, (2012) define leadership styles as a "...pattern of behaviors engaged in by the leader when dealing with employees". For instance, Nanjundeswaraswamy and Swamy (2014) noted that the leadership style of a leader would affect the organizational commitments and work satisfaction to improve work performance. Furthermore, while leadership styles influence the achievement of set objectives of projects, certain factors inspire these leadership styles. Goh Yuan Sheng and Soutar (2005) argue that leadership styles are significantly influenced by the leaders' immediate and external environment. Thus, the situation is significant, whereas Podsakoff, MacKenzie, Moorman, and Fetter, (1990) suggest that the behavior of leaders will affect their leadership styles. Also, Amanchukwu, Stanley, and Ololube, (2015) postulate that the degree of interaction (communication) and leaders' personality affected their leadership style.

Nanjundeswaraswamy and Swamy (2014), in their study on leadership styles, reported on two types of leadership styles, namely the transformational and transactional leadership styles. Constantin (2011) identified three broad leadership styles. He further outlines them as autocratic or authoritative, participatory or democratic and laissez-faire or free reign leadership styles. Val and Kemp (2012) also classified leadership styles into three main types, namely the autocratic (authoritarian), democratic (participative) and laissez-faire (abdicratic) leadership styles in studying the roles played by a leader in affecting the dynamics of a large group while partaking in a field expedition. Amanchukwu et al. (2015) also studied the relevance of leadership in educational management and discussed six leadership types which include transactional, autocratic, democratic, laissez-faire, bureaucratic and charismatic leadership styles. The study, therefore, will discuss seven unique leadership styles which have been identified and may be relevant to the M&E of construction project delivery.

8.3.1 Transformational leadership style

The transactional leadership style is a vision-mission-centered approach to leadership towards project success. According to Sultana, Darun and Yao (2015), the transactional leadership style entails working and providing direction to transform subordinates' interest towards team or group (collective) interest (the mission of the project). The transformational leader therefore inspires, convinces and guides team members to achieve the outcome (Robbins, 1996), hence moving subordinates from the thinking of existence to the thinking of growth, achievement and development (Bass et al., 1990). The leader's vision is guided by the objectives and desired outcome of the project being undertaken and therefore he or she formulates strategies to appeal and make the vision attractive through awareness of the vision to improve followers' commitment that will lead to achieving such desired outcomes (Sultana et al., 2015). Transformational leadership is often defined in terms of the impact leaders have on followers; followers see leaders as trusted, admired, loyal and have respect for the leader and are also motivated to do more than they initially anticipated to do (Goh et al., 2005). Bass and Avolio (1990) further posit that transformational leaders exhibit four basic characteristics, namely idealized influence, individual consideration, intellectual stimulation and inspirational (Popa, 2012).

8.3.2 Transactional leadership style

The transactional leadership style is described as a mutually dependent relationship between leaders and subordinates (Kuhnert & Lewis, 1997). Leaders acknowledge and reward subordinates in exchange for work done. Thus, followers are compensated for achieving specific goals and performance criteria (Amanchukwu et al., 2015; Nanjundeswaraswamy & Swamy, 2014). Two levels of transactional leadership styles have been reported by Kuhnert and Lewis (1997). These they outline as high-quality and low-quality transactional leadership styles. A transactional leadership style which involves support and exchange of emotional resources, including the reward, is regarded as high-quality which is most likely to keep subordinates continuing working whereas low-quality transactions are strictly based on the contractual agreement (Kuhnert & Lewis, 1997). However, it is argued by Bass et al. (1990) that subordinates are motivated through contingency rewards, management by exception, rule enforcement and corrective actions undertaken via close monitoring activities (Judge, Fluegge Woolf, Hurst & Livingston, 2006; Koech & Namusonge, 2012).

8.3.3 Autocratic/authoritative leadership style

According to Amanchukwu et al. (2015), autocratic leadership exhibits extreme characteristics of the transactional leadership style. Complete control over staff and team members is vested in the leader with little or no opportunity for staff to make suggestions, notwithstanding who stands to benefit (Amanchukwu et al., 2015; Constantin, 2011). Final decisions are taken by a leader without any consultation; subordinates are not trusted and there are no delegatory roles (Khan et al., 2015). Influencing subordinates to undertake any activity is done with intimidation and a sense of fear; this results in poor performance due to subordinates' voluntarily resigning from work or seeking a transfer, lateness to work or reduction in work output (Akor, 2014). The leadership style is, however, effective in emergency situations where there is no time to discuss and agree. Constantin (2011) further argues that smaller groups and capable teams working under this type of leadership styles are motivated to work effectively and achieve results. The motivation of subordinates is achieved through the establishment of structured sets of rewards and punishment (Khan et al., 2015).

8.3.4 Democratic or participatory leadership style

While it appears difficult to propose a universal definition for democratic leadership styles, Gastil (1994) re-echoes the basic elements consistent with democratic principles to describe it. These principles included self-determination, equal participation, inclusiveness and deliberation (Choi, 2007). Whereas a democratic leadership style may be time-consuming since the views of followers are incorporated in the decision-making process through lengthy debates, it is argued that it encourages participation, member satisfaction and increased follower productivity (Choi, 2007). Therefore, due to the inclusion of all subordinates in the decision-making process, a highly skilled and knowledgeable following is appropriate in such a work environment. Ray and Ray (2012) posit that democratic leaders serve as moderators and facilitators while encouraging followers to contribute, thereby empowering them in the process. Consequently, in exercising the final decision, a combination of relevant and knowledgeable contributions from followers is imperative.

8.3.5 Laissez-faire leadership style

Laissez-faire is the type of leadership style with little or no involvement of the leader in any activity or decision making. It describes a situation where leaders abdicate their duties and responsibilities to subordinates (Amanchukwu et al., 2015). Power to make decisions is abdicated to subordinates (Chaudhry & Javed, 2012) while providing resources and advice when needed. Subordinates run the day-to-day activities of the work by establishing their own goals for the work (Constantin, 2011). Laissez-faire leadership may prove effective when performance monitoring is undertaken from afar and regular feedback provided; it could possibly lead to job satisfaction and increased productivity due to the autonomy of subordinates. On the other hand, it can pose to be the worse form of leadership style if subordinates are less skilled, knowledgeable and motivated to do the work.

8.3.6 Bureaucratic leadership style

The bureaucratic leadership is among the oldest forms of leadership style described in literature. The bureaucratic leadership style can somehow be linked to the transactional leadership style as it rigorously follows the rules, regulations and adheres to authority (Amanchukwu et al., 2015). Characteristics of bureaucratic leaders as reported by Rouzbahani, Alibakhshi, Ataie, Koulivand, and Goudarzi, (2013) include the imposition of strict and systematic discipline on followers regarding workplace conduct and the ability to conform to office rules ensures followers promotion to higher ranks. Further, leaders are empowered through the office and position they occupy and, finally, followers obey leaders because of the authority bestowed on them as part of their position in the company (Rouzbahani et al., 2013). Based on organizational policies, guidelines and processes, bureaucratic leaders instruct followers on what to do on a daily basis with little or no freedom (Amanchukwu et al., 2015). The bureaucratic leadership type is seen as appropriate in a work environment where health and safety risks and measures are to be safeguarded (Amanchukwu et al., 2015) as well as a work environment involving routine tasks (Schaefer, 2005). However, the bureaucratic leadership style is known for suppressing creativity and innovation which is a major setback for the bureaucratic leadership type (Santrock, 2007).

8.3.7 Charismatic leadership style

The charismatic leadership style is very much like transformational leadership. Perhaps the only difference between the two leadership styles is seen in the approach towards subordinates' interests. France (2008) postulates that charismatic leaders transform subordinates' interest to match that of their own interests whereas transformational leaders seek the transformation of subordinates' interest to be in line with the group or teams' interests. He inspires the subordinate to share in his vision to accomplish the task; this, therefore, places more emphasis on the leader than on employees, creating a feeling of all-knowing by the leader. This, however, is dire to the organization and the work when leaders have to quit (Amanchukwu et. al., 2015). Four key distinct characteristics have been reported of charismatic leadership styles, namely the willingness to take the risk to achieve the vision, exhibiting sensitivity to followers' needs, demonstrating novel behavior and possessing and articulating a clear vision (Judge et al., 2006).

8.4 Leadership theories: An evolution tree approach

One of the most complex phenomena to which organizational and psychological research has been applied is leadership. Whereas the term "leadership" became prominent in the late 1700s, the term "leader" was well-known and conceptualized as early as the 1300s before the biblical times (Stogdill, 1974). Nonetheless, not until the twentieth century has scientific research on the topic taken centerstage (Bass, 1981). Up until now, a proliferation of studies on leadership has revealed several leadership theories and models that have evolved over the years to define and put the leadership styles into perspective. This subsection, therefore, chronicles the development of leadership theories (presented graphically in Figure 8.1) over ten leadership eras and periods. The model recognizes the order in which leadership theories developed. Each new era presents a higher stage of development of the previous era.

8.4.1 Personality era

The personality era marks the beginning of the leadership development process and presents two periods, namely the great man period (Great Man Theory) and the trait period (Trait Theory). The great man period suggested that imitating the personality and behaviors of great men (and sometimes women) in the history of the world would make one a strong leader (Gallon, 1869). In earlier studies, leadership was equated to personality (Bowden, 1927) and sometimes inheritance (Jennings, 1960). The great man theory therefore portrayed leaders as heroic and mythic. This belief became unpopular since it was clear there was difficulty in

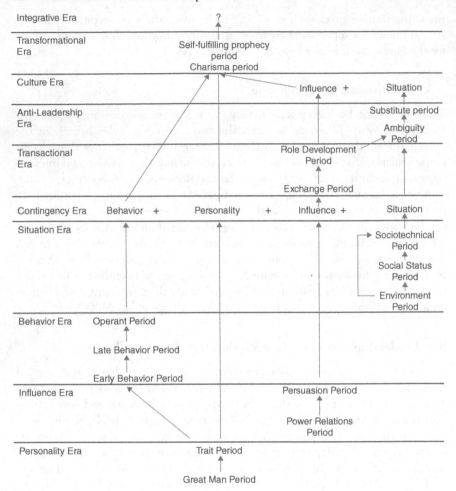

Figure 8.1 Evolution tree of leadership theory.

Source: Van Seters & Field, 1990

imitating varied personalities. The trait period came with little advancement in the development of leadership theories (King, 1990). Some key traits were developed such as the desire to lead, drive, integrity, self-confidence, intelligence and relevant job knowledge which, if adopted, would enhance the leadership potential of persons (Maslanka, 2004). The focus is on what makes a leader. Similar to the great man theory, the trait theory failed when studies revealed there were no single or sets of traits that were consistent with effective leadership (Maslanka, 2004; Van Seters & Field, 1990). The search for a new explanation regarding effective leadership was therefore necessary when there were inconsistencies between leadership traits and leadership effectiveness (Amanchukwu et al., 2015). Then the influence era followed.

8.4.2 Influence era

The influence era also saw two periods, namely the power relations and the persuasion periods. This era recognized that the relationship between individuals defines leadership. Therefore, it sought to emphasize the power and influence of persons in a relationship. The power relation period implied that the degree of leadership effectiveness lies in the source, power and utilization of such power. Evidence of the influence of power is seen in authoritative leadership. However, French (1956) argues how ineffective they are in current leadership approaches. The second period of the influence era was the persuasion period which frowns on intimidation, force, authoritarian and bullying leaders. This era, however, recognized that in a leader-member relationship, the leader rules; thus, the leader's dominance approach.

8.4.3 Behavior era

An entirely new focus on leadership described effective leaders by what they do. Studies in this era emphasized the behavior pattern of leaders as well as the differences between an effective and an ineffective leader (Yukl, 1971). This era has been described as a major advancement in the leadership theory for two reasons: it enjoyed the strong empirical support and its ease of implementation (Fleishman & Harris, 1962). Three periods were included in the behavior era; the early behavior period, late behavior period and the operant period. The early behavior period focused on advancing the personality trait by developing behavioral traits. Leaders' interest in accomplishing the task and their concern for individual group cohesion were the two core leadership behavior traits identified with the early behavior period by the Ohio State and Michigan (Griffin, Skivington & Moorhead, 1987). These behavioral features are seen exhibited by charismatic leaders. What the late behavioral period did was to adapt to the early behavioral period theories for management application. The theories X and Y received much attention.

McGregor (1966) explained that whereas theory X suggests that subordinates' employees were passive, disliked work, performed only under guidance and supervision, were not responsible and required intimidation to perform and also enjoyed financial motivation, theory Y described subordinates as inherently motivated, responsible for working, enjoyed work which they regarded as naturally part of life, required little or no guidance to work and performed effectively and efficiently in a conducive work environment. The operant period discussed the behavior of leaders as a reflection of subordinates. Thus, subordinates behaved like their leaders. Kerr and Schriesheim (1974) advanced that though the behavior era was well researched, there existed mixed empirical evidence to support it. This era entrenched the behavior theory which attributes great leadership as being born and not made (Amanchukwu et al., 2015), suggesting people can learn to become great leaders through training.

8.4.4 Situation era

The situation era significantly advanced the frontiers of leadership by looking beyond personal and inherent factors of leaders and subordinates' characteristics. Great leaders are those who can adapt to every situation and stay on top of their duties. Three periods emerged within the situation era; the environmental period which indicates that a great leader is the one who is always at the right place at the right time (King 1990). It agrees that no specific individual is vested with leadership, but rather other individuals can rise to the occasion and lead when it matters (Hook, 1943). The practical support gained by this approach saw the proposition by McCall (1977) to develop more environmental factors for use in the context of leadership. The second period was the social status period and focused on the social aspects of a situation. It suggested that the social status of individuals influenced their leadership style and made them effective or otherwise. The third period was the nontechnical period. This period combined the environmental and social status, drawing on its strength and weakness in describing an effective leader.

8.4.5 Contingency era

The contingency era saw several theories which acknowledged no single leadership style as appropriate for all situations; therefore multiple factors, including personality, behavior, influence and situational, defined a great leader (Van Seters & Field, 1990). Three major theories were noted within this era, namely the contingency theory, the path-goal theory and the normative theory (King, 1990). The contingency theory suggests that leaders are placed in suitable situations to match their leadership styles or are trained to change the situation to match their style (Fiedler, 1967). Similarly, contingency theory contends that there exist several ways of leading and organizing based on the relationship between two factors, namely the leadership style and the situational suitability. Therefore, a leadership style which may be effective in one situation may fail in another situation; hence contingent on various internal and external factors (Fiedler, 1967).

The path-goal theory is reported by Talal Ratyan, Khalaf and Rasli (2013) as the most effective contingency approach to leadership. The path-goal theory focuses on providing an enabling condition for subordinates to succeed rather than focusing on situation and leaders' behavior (House, 1971). Also, the path-goal theory concentrates on motivating factors of subordinates which may significantly influence the task (Talal Ratyan et al., 2013). The normative theory, on the other hand, submits that leaders could change their behavior to match particular situations to succeed and so effective leadership is measured by how leaders adapt to situations by changing their behavior (Vroom & Jago, 2007). Ayman and Korabik (2010) further submit that the normative leadership theory is situationally centered and assume that leadership responds to situational determinants.

8.4.6 Transactional era

Role differentiation and social interaction were the underpinnings of this leadership era (King, 1990). This era looked at leadership beyond mere personal or situational factors or characteristics but rather the relationship between leaders and subordinates. Roles were clearly and mutually set and rewards and punishments attached to performance (Cherry, 2016). The transactional era also saw two periods: the exchange period and the role development period. In the exchange period, one could only be said to be a great leader, not by personal or behavioral qualities, but through the transactions between leaders and subordinates which establish a relationship. Great leaders would therefore be acknowledged by group members or subordinates as a great leader based on the strength of the established relationship (Bass, 1981). Among the theories propounded during the exchange period were the reciprocal influence approach, the vertical dyad linkage theory and the leader-member exchange theory. During the role development period, the emphasis was placed on the role of leaders and subordinates. Characteristics of this period were the development of the social exchange theory and the role-making model. A controversy, however, affected this period, suggesting leadership did not lie only in leaders but also in subordinates. This generated a great deal of confusion, sending researchers back to their roots to redefine the domain of leadership (King, 1990).

8.4.7 Anti-leadership era

The anti-leadership era emerged after a series of experiments revealed that there existed nothing like a leadership concept. In that sense of the no-leadership concept, the anti-leadership era (the belief of no leadership) was characterized by two periods, namely the ambiguity period and the substitute period. The ambiguity period posits that leadership was only a perception held by people and did not exist (King, 1990). It was reported that the perception of leadership was a symbol which had no consequence on performance (Pfeffer, 1977). Further, Miner (1975) postulated an abandonment of the concept of leadership whereas Meindl, Ehrlich, and Dukerich, (1962) intimated that changes in an organization described by the concept of leadership were not understood. The **substitute period** evolved from the attempt by the situational era to establish an alternative to the concept of leadership. King (1990) citing Kerr and Jermier (1978) indicated the task of subordinates as well as the organization prevented leadership from affecting the performance of subordinates. An inevitable limitation of leadership influence on the subordinates' performance existed, thereby suggesting the introduction of substitutes and leadership neutralizers to be incorporated by leaders to ensure their continued presence and influence.

8.4.8 Culture era

The focus of leadership shifted from an increase of the volume of work accomplished to describe an increased quality of work. As reported by Van Seters and

Field (1990), the macro view of leadership as reported by Pascale and Athos (1981) was evident in the 7-S framework. Similarly, the search for excellence and the theory Z describes employee involvement as key to performance and shared responsibility while performing better in a trusted and cooperative environment (Peters & Waterman, 1982; Ouchi & Jaeger, 1978; Ouchi, 1981). As an extension of the leader substitute period of the anti-leadership era, it is hoped that employees would lead themselves if leaders create a strong culture in the organization (King, 1990). Leadership was therefore automatic with an established strong culture and would require formal leadership only when the existing culture is changed. Schein, (1985) informed that organizations could run without the presence of leadership, except at the initial planning stage.

8.4.9 Transformational era

The most recent and promising era of leadership can be said to be the transformational era of leadership evolution. It is an improvement over all the eras discussed above and it is based on external motivation as opposed to the intrinsic motivation of earlier eras. The transformational era suggests that leaders must be proactive in thinking, be more radical and open to innovative ideas. A true leader influences and encourages enthusiasm and commitment of subordinates rather than reluctant obedience and indifferent (Yukl, 1971). Two periods are recognized in this era, namely the charisma period and the self-prophecy period. Building on the culture era, leadership during the charisma period is seen as a collective action rather than a one-person show (Roberts, 1985) towards transforming vision and giving a new and stronger sense of meaning and purpose to all who share in the mission and vision.

This period advocates for leaders who formulate and empower subordinates to carry out the vision. The charismatic leadership theory, also known as the relationship theory, emerged at this time. The philosophy underpinning the charismatic theory is its holistic nature in which leaders' influence, traits, behavior and situational factors are combined to enhance the acceptance of the principles of leaders. The self-fulfilling prophecy period is based on the self-fulfilling prophecy phenomenon (Field, 1989). This advanced that leadership occurred from leaders to subordinates and vice versa. The model of a self-fulfilling prophecy was therefore developed within this period (Eden, 1984). Bass (1985), in his book *Leadership and Performance Beyond Expectations*, stressed the selection of leaders based on the leader's ability to accomplish the task, maintain a strategic focus and facilitate group cohesion.

8.5 Barriers to effective monitoring and evaluation leadership

The Chartered Institute of Building (CIOB) (2008), according to Adair (1973), describes the function of leadership in construction as achieving the task, building a team and developing individuals. These functions, however, have been described as the project implementation barriers confronting leadership in the

Malaysian construction industry, according to Nasaruddin and Rahman (2017), and need to be addressed urgently. These features summarize the role of leadership in the M&E of construction projects. The actions of leadership in planning, coordination and execution beget project success. Undertaking these roles effectively however has received challenges resulting in ineffective leadership in the M&E process in achieving project success. This section therefore discusses some barriers that have caused leadership to be ineffective in ensuring effective M&E.

Owing to the complexity and multidisciplinary nature of the construction industry, project managers face various challenges in their daily activities. Ofori and Toor (2012) argue for considerable attention towards leadership in the construction industry due to the feature and processes of projects in the construction industry. It can also be said that as a result of the multiplicity of stakeholders with varying interests, skilled and unskilled personnel involved in project delivery, leadership in the M&E process is more essential to guarantee project success. CIOB (2008) in a study on leadership in the construction industry identified two main challenges that prevent leaders from reaching their full potential in delivering their leadership mandate in project delivery. These challenges observed are the lack of opportunity for project managers to improve their leadership skills and, secondly, a poor organizational culture towards leadership and the over-concentration on the technical competence (Archer, Verster & Zulch, 2010). Archer et al. (2010) revealed the lack of experience and skill as reasons for the poor leadership of project managers. This they attribute to insufficient education and lack of training which requires immediate attention.

The male dominance of the construction industry poses a challenge for women in leadership. For instance, in 1996, the chairperson of the CIOB, Prof. Michael Romans threatened to resign if the institute's attitude towards women is not changed (Thurairajah, Amaratunga & Haigh, 2007). However, Hall-Taylor, (1997) informed that the negative perception, personality and self-motivation of women leaders is self-inflicted and poses as a barrier to women leaders in a male-dominated environment – even though studies have shown that gender makes no difference to effective leadership in the construction industry. Lack of confidence and assertiveness to aspire to managerial and leadership positions (Omar & Ogenyi, 2004), failure to undertake training and develop managerial skills (Thurairajah et al., 2007), choice of field of education and multiple commitments (Fielden, Davidson, Gale & Davey, 2000), challenge the leadership of women in the M&E of construction project delivery. In a related study which discussed barriers to female leadership, it is reported that the gender of women greatly affected their leadership potential in organizations (Titrek, Bayrakci & Gunes, 2014). According to Titrek et al. (2014), this adverse effect of the female gender towards leadership is influenced by their low educational levels and traditional responsibilities such as caring for children and housework.

According to Bikitsha, Mamafha and Ngomane (2014), poor relationships and ineffective communication significantly contribute to the ineffectiveness of the leadership process in construction project M&E, indicating that projects are likely to fail as a result of the weak relationship between leader and other project stakeholders or project participants (Meng, 2011). It is reported that effective leadership is effective communication (Luthra & Dahiya, 2015). Hence, a leader's failure to communicate and coordinate the M&E process effectively negatively affects project performance (Loosemore & Lee, 2002). Similarly, the inevitable conflict in construction (Kumaraswamy, 1997; Yiu & Cheung, 2006) due to the varied interests of stakeholders and the project team and its effect on leadership has been widely reported and therefore requires effective communication by project leaders (Mitkus & Mitkus, 2014).

Exercising a democratic leadership style where extensive consultation among stakeholders and the project team is required in an emergency situation may result in project failure; thus, the project may be delayed for some decisions to be taken collectively which has a consequential impact on other components of the project. For leadership to be efficient and surmount situational challenges and vice versa, a multi-adaptive leadership style is essential. Thus, an effective leader is one who can adjust his or her leadership method to achieve the desired objective in a difficult situation (Luthra & Dahiya, 2015). The company culture of organizations comprises the norms, values, beliefs and assumptions of the organization and may differ from one organization to the other. Hence, ignorance of project managers and leaders about such cultures may cause leaders to be ineffective. However, leaders' understanding of the organizational culture will inform their leadership approach and will ensure high performance of team members. Cultural backgrounds and unique understanding levels of team members make leadership communication a challenge (Luthra & Dahiya, 2015).

8.6 Achieving effective monitoring and evaluation leadership

Achieving complete leadership effectiveness is not possible, particularly in the construction industry owing to the complexities discussed in the preceding chapters. According to Luthra and Dahiya (2015), effective leadership is all about communicating effectively. Perhaps this is correct since communication in construction project M&E requires that leaders manage and coordinate project resources (human, financial and material) and processes for efficiency. Effective leadership therefore requires excellent communication skills (Luthra & Dahiya, 2015). It is further asserted by Luthra and Dahiya (2015) that a clear set of values and indoctrinating these values into project team members will get team members to follow the leader to achieve the vision, thereby making him or her an effective leader. The World Health Organization (WHO) (2008) reports that good management skills, that is, clarity of task and purpose, good organizational skills and good delegatory skills of leaders, are central for effective leadership. Leaders must be knowledgeable (Popa, 2012). Complementing the managerial skills of leaders, leaders with technical knowledge on M&E and project implementation in the construction industry is necessary.

According to Kolzow (2014), leadership competencies such as the ability to inspire others to deliver task, the collaborative orientation of leaders, exhibiting intelligence and learning agility, practicing accountability and exerting real judgment cannot be overlooked if leadership effectiveness is to be achieved. He further stressed that these competencies are not inherent or inborn; therefore, through a concerted effort by leaders, one can acquire these competencies for ensuring positive leadership (Kolzow, 2014). Studies on the influence of leadership style on the project performance have been conducted (Iqbal, Anwar & Haider, 2015). Iqbal et al. (2015) studied the relationship between several leadership styles on project performance, namely autocratic, democratic and participative leadership style and concluded that there were differences in project performance regarding each leadership style studied (Iqbal et al., 2015), suggesting that the positive or the right leadership style is significant to accomplish effective leadership.

The study of leadership theories has underscored the far-reaching demands on leaders to achieve organizations' performance. The many leadership styles indicate the difficulty in achieving effectiveness in the dynamic project environment such as the construction industry (Bikitsha et al., 2014). A combination of several styles is needed. Indeed, Batool (2013) found that effective leaders combined more than four leadership styles and substituted each one for a particular leadership situation. Accordingly, the M&E process in the construction industry is emotionally charged, considering the varied stakeholders involved in the M&E implementation. In addressing this, studies have reported on the need for leaders to possess social and emotional intelligence to deliver effective leadership (Batool, 2013; Bikitsha et al., 2014).

In accomplishing effective leadership in the M&E of construction projects, leaders of construction projects must demonstrate effective communication, managerial skills, technical knowledge, leadership competencies and emotional intelligence.

Summary

This chapter debated effective leadership in the M&E of construction projects as a critical construct in the integrated M&E model for effective construction project delivery. A combination of both old and new leadership theories such as the traits theory and transformational leadership theories respectively were captured as the grounded theories of effective leadership. This was systematically done through the various evolution eras of leadership theories. Also emphasized in this chapter are the types and style of leadership, the barriers to effective leadership and, finally, how to achieve effective leadership in M&E of projects. Most importantly, strategies for achieving leadership in M&E were suggested. The next chapter presents a global assessment view of M&E practices in the UK and Australia.

References

Adair, J. (1973). *Action-centred leadership*. London, New York: McGraw-Hill, Inc.

Akor, P. U. (2014). Influence of autocratic leadership style on the job performance of academic librarians in Benue State. *Journal of Educational and Social Research*, 4(7), pp. 148–152, doi:10.5901/jesr.2014.v4n7p148

Amanchukwu, R. N., Stanley, G. J. & Ololube, N. P. (2015). A review of leadership theories, principles and styles and their relevance to educational management. *Management*, 5(1), pp. 6–14.

Archer, M. M., Verster, J. J. & Zulch, B. G. (2010). Leadership in construction project management: Ignorance and challenges. In: *Proceedings 5th Built Environment Conference*. Presented at the Built Environment Conference, Durban, South Africa.

Ayman, R. & Korabik, K. (2010). Leadership: Why gender and culture matter. *American Psychologist*, 65(3), pp. 157–170, doi:10.1037/a0018806

Bass, B. M. (1981). *Stogdill's handbook of leadership: A survey of theory and research. (Revised and expanded version.)*. New York: Free Press.

Bass, B. M. (1985). *Leadership and performance beyond expectations*. New York: Free Press.

Bass, B. M. (1990). *Bass and Stogdill's handbook of leadership*. 3rd Edition. New York: The Free Press.

Bass, B. M. & Avolio, B. J. (1990). Developing transformational leadership: 1992 and beyond. *Journal of European Industrial Training*, 14(5), doi:10.1108/03090599010135122

Batool, B. F. (2013). Emotional intelligence and effective leadership. *Journal of Business Studies Quarterly*, 4(3), p. 84.

Bennis, W. G. (1989). *On becoming a leader*. Reading, MA: Addison-Wesley.

Bhatti, N., Maitlo, G. M., Shaikh, N., Hashmi, M. A. & Shaikh, F. M. (2012). The impact of autocratic and democratic leadership style on job satisfaction. *International Business Research*, 5(2), doi:10.5539/ibr.v5n2p192

Bikitsha, L., Mamafha, K. & Ngomane, N. (2014). Understanding the use of emotional intelligence during the project leadership process: A case of project managers. *Journal of Leadership and Management Studies*, 1(1), pp. 5–16.

Bowden, A. O. (1927). A study on the personality of student leadership in the United States. *Journal of Abnormal Social Psychology*, 21, pp. 149–160.

Burns, J. M. (1978). *Leadership*. New York: Harper and Row.

Chartered Institute of Building (2008). *Leadership in the construction industry*. Berkshire, UK: The Chartered Institute of Building.

Chaudhry, A. Q. & Javed, H. (2012). Impact of transactional and laissez faire leadership style on motivation. *International Journal of Business and Social Science*, 3(7). pp 258–264.

Chemers, M. M. (1997). *An integrative theory of leadership*. Mahwah, NJ: Lawrence Erlbaum.

Cherry, K. (2016). *The major leadership theories: The eight major theories of leadership*. Available online at: https://www.verywell.com/leadership-theories-2795323 [Accessed 25 August 2017].

Choi, S. (2007). Democratic leadership: The lessons of exemplary models for democratic governance. *International Journal of Leadership Studies*, 2(3), pp. 243–262.

Constantin, D. (2011). Leadership Styles. *Defense resources management in the 21st century*. 6th International Scientific Conference. December 2nd–3rd, 2011, Brasov, Romania.

Eden, D. (1984). Self-fulfilling prophecy as a management tool: Harnessing Pygmalion. *Academy of Management Review*, 9, pp. 64–73.

Field, R. H. G (1989). The Sell-Fulfilling Prophecy Leader: Achieving the Metharme Effect, *Journal of Management Studies*, 26,151–75.

Fiedler, F. E. (1967). *A theory of leadership effectiveness*. New York: McGraw-Hill.

Fielden, S. L., Davidson, M. J., Gale, A. W., & Davey, C. L. (2000). Women in construction: The untapped resource. *Construction management and economics*, 18(1), 113–121. https://doi.org/10.1080/014461900371004

Fleishman, E. A. & Harris, E. F. (1962). Patterns of leadership behaviour related to employee grievances and turn-over. *Personnel Psychology*, 15, pp. 43–56.

France, S. H. (2008). *Leadership theories: Toward a relational model*. Retrospective Exam, EXD-66909 for the Ad hoc doctoral programme of Administrative Sciences, Université Laval, Québec.

French, J. R. (1956). A formal theory of social power. *Psychological Review*, 63, pp. 181–194.

Gallon, F. (1869). *Hereditary genius*. New York: Appleton.

Gardner, J. W. (1990). *On Leadership*, Free Press, New York, NY.

Gastil, J. (1994). A definition and illustration of democratic leadership. *Human Relations*, 47(8), pp. 953–975.

Goh Yuan Sheng, V. & Soutar, G. N. (2005). The role of ethical behaviours in the relations between leadership styles and job performance. *ANZMAC 2005 Conference: Corporate Responsibility*.

Griffin, R. W., Skivington, K. D. & Moorhead, G. (1987). Symbolic and international perspectives on leadership: An integrative framework. *Human Relations*, 40, pp. 199–218.

Hall-Taylor, B. (1997). The construction of women's management skills and the marginalization of women in senior management. *Women in Management Review*, 12(7), pp. 255–263, doi:10.1108/09649429710181225

Hook, S. (1943). *The hero in history*. New York: John Day.

House, R. J. (1971). A path-goal theory of leader effectiveness. *Administrative Science Quarterly*, 16, pp. 321–338.

Iqbal, N., Anwar, S. & Haider, N. (2015). Effect of leadership style on employee performance. *Arabian Journal of Business and Management Review*, 5(5). pp 1–6.

Jennings, E. E. (1960). *An anatomy of leadership*. New York: Harper.

Judge, T. A., Fluegge Woolf, E., Hurst, C. & Livingston, B. (2006). Charismatic and transformational leadership: A review and an agenda for future research. *Zeitschrift für Arbeits- und Organisationspsychologie A&O*, 50(4), pp. 203–214, doi:10.1026/0932-4089.50.4.203

Kerr, S. & Schriesheim, S. (1974). Consideration, initiating structure, and organizational criteria: An update of Korman's 1966 review. *Personnel Psychology*, 27, pp. 555–568.

Kerr, S. & Jermier, J. M. (1978). Substitutes for leadership – Their meaning and measurement. *Organizational Behaviour and Human Performance*, 18, pp. 329–345.

Khan, M. S., Khan, I., Qureshi, Q. A., Ismail, H. M., Rauf, H., Latif, A. & Tahir, M. (2015). The styles of leadership: A critical review. *Public Policy and Administration Research*, 5(3), pp. 87–92.

King, A. S. (1990). Evolution of leadership theory. *Vikalpa*, 15(2), pp. 43–56.

Koech, P. M. & Namusonge, G. (2012). The effect of leadership styles on organizational performance at state corporations in Kenya. *International Journal of Business and Commerce*, 2(1), pp. 1–12.

Kolzow, D. R. (2014). Leading from within: Building organizational leadership capacity. *International Economic Development Council*, pp. 1–314.

Kuhnert, K. W. & Lewis, P. (1997). Transactional and transformational leadership: A constructive/developmental analysis. *Academy of Management Review*, 12(2), pp. 648–657.

Kumaraswamy, M. M. (1997). Conflicts, claims and disputes in construction. *Engineering, Construction and Architectural Management*, 4(2), pp. 95–111, doi:10.1108/eb021042

Loosemore, M. & Lee, P. (2002). Communication problems with ethnic minorities in the construction industry. *International Journal of Project Management*, 20(7), pp. 517–524.

Luthra, A. & Dahiya, R. (2015). Effective leadership is all about communicating effectively: Connecting leadership and communication. *International Journal of Management & Business Studies*, 5(3), pp. 43–48.

Maslanka, A. M. (2004). *Evolution of leadership theories*. Allendale, Michigan: Grand Valley State University.

McCall, M. W. (1977). *Leadership: Where else can we go?* Durham: Duke University Press.

McGregor, D. (1966). *Leadership and motivation.* Cambridge: MIT Press.

Meindl, J. R., Ehrlich, S. B. & Dukerich, J. M. (1962). The romance of leadership. *Administrative Science Quarterly*, 30, pp. 78–102.

Meng, X. (2011). The effect of relationship management on project performance in construction. *International Journal of Project Management*, 30(2), pp. 188–198, doi:10.1016/j.ijproman.2011.04.002

Miner, J. B. (1975). *The uncertain future of the leadership concept: An overview.* Leadership Frontiers. Kent: Kent State University Press.

Mitkus, S. & Mitkus, T. (2014). Causes of conflicts in a construction industry: A communicational approach. *Procedia – Social and Behavioral Sciences*, 110, pp. 777–786, doi:10.1016/j.sbspro.2013.12.922

Nanjundeswaraswamy, T. S. & Swamy D. R. (2014) Leadership styles. *Advances In Management.* 7(2), pp 57–62.

Nasaruddin, N. A. N. & Rahman, I. A. (2017). Exploratory study on Malaysia construction leadership. In: *MATEC Web of Conferences*, EDP Sciences, 1–6.

Ofori, G. & Toor, S. R. (2012). Leadership and construction industry development in developing countries. *Journal of Construction in Developing Countries*, Supp. 1, pp. 1–21.

Omar, O. & Ogenyi, V. (2004). A qualitative evaluation of women as managers in the Nigerian Civil Service. *International Journal of Public Sector Management*, 17(4), pp. 360–373, doi:10.1108/09513550410539839

Ouchi, W. G. (1981). *Theory Z: How American business can meet the Japanese challenge.* Reading: Addison-Wesley.

Ouchi, W. G. & Jaeger, A. M. (1978). Type Z organization: Stability in the midst of mobility. *Academy of Management Review*, 3, pp. 305–314.

Panthi, K., Farooqui, R. U., & Ahmed, S. M. (2008). An investigation of the leadership style of construction managers in South Florida. *Journal of Construction Management and Economics*, 11(4), 455–565.

Pascale, R. T. & Athos, A. G. (1981). *The art of Japanese management: Application for American executives.* New York: Warner Books.

Peters, T. J. & Waterman, R. H. (1982). *In search of excellence: Lessons from America's best-run companies.* New York: Warner Books.

Pfeffer, J. (1977). The ambiguity of leadership. *Academy of Management Review*, 2, pp. 104–112.

Podsakoff, P. M., MacKenzie, S. B., Moorman, R. H. & Fetter, R. (1990). Transformational leader behaviors and their effects on followers' trust in leader, satisfaction, and organizational citizenship behaviors. *The Leadership Quarterly*, 1(2), pp. 107–142.

Popa, B. M. (2012). The relationship between leadership effectiveness and organizational performance. *Journal of Defense Resources Management*, 3, 1(4), pp. 123–126.

Ray, S. & Ray, I. A. (2012). Understanding democratic leadership: Some key issues and perception with reference to India's freedom movement. *Afro-Asian Journal of Social Sciences*, 3(3.1), pp. 1–26.

Robbins, S. P. (1996). *Organizational behavior: Concepts, controversies, applications.* Englewood Cliffs, NJ: Prentice Hall.

Roberts, N. C. (1985). Transforming leadership: A process of collective action. *Human Relations*, 38, pp. 1023–1046.

Rouzbahani, M. T., Alibakhshi, D., Ataie, S., Koulivand, A. & Goudarzi, M. M. (2013). The relationship between bureaucratic leadership style (task-oriented) and customer relationship management (CRM). *Journal of Basic and Applied Scientific Research*, 3(2), pp. 1292–1296.

Santrock, J. W. (2007). *A topical approach to life-span development.* New York, NY: McGraw-Hill.

Schaefer, R. T. (2005). *Sociology.* 9th edition. New York, NY: McGraw-Hill.

Schein, E. H. (1985). *Organizational culture and leadership.* San Francisco: Jossey-Bass.

Sharma, M. K. & Jain, S. (2013). Leadership management: Principles, models and theories. *Global Journal of Management and Business Studies,* 3(3), pp. 309–318.

Stogdill, R. M. (1974). *Handbook of leadership.* New York: Free Press.

Sultana, U. S., Darun, M. R. & Yao, L. (2015). Transactional or transformational Leadership: Which works best for now? *International Journal of Industrial Management.* pp 1–8.

Talal Ratyan, A., Khalaf, B. & Rasli, A. (2013). Overview of path-goal leadership theory. *Jurnal Teknologi,* 64(2), doi:10.11113/jt.v64.2235

Thurairajah, N., Amaratunga, D. & Haigh, R. (2007). Confronting barriers to women in leadership positions: A study on construction industry. *Presented at the 7th International Postgraduate Conference in the Built and Human Environment,* Salford Quays, UK.

Titrek, O., Bayrakci, M. & Gunes, D. Z. (2014). Barriers to women's leadership in Turkey. *Anthropologist,* 18(1), pp. 135–144.

Van Seters, D. A. & Field, R. H. G. (1990). The evolution of leadership theory. *Journal of Organizational Change Management,* 3(3), pp. 29–45, doi:10.1108/09534819010142139.

Val, C. & Kemp, J. (2012). Leadership styles. *Pathways: The Ontario Journal of Outdoor Education,* 24(3), pp. 28–31.

Vroom, V. H. & Jago, A. G. (2007). The role of the situation in leadership. *American Psychologist,* 62(1), pp. 17–24, doi:10.1037/0003-066X.62.1.17

World Health Organization (WHO). (2008). Leadership and management. In: *Operations manual for staff at primary health care centers.* Switzerland: WHO Press, pp. 264–281.

Yiu, K. T. W. & Cheung, S. O. (2006). A catastrophe model of construction conflict behavior. *Building and Environment,* 41(4), pp. 438–447, doi:10.1016/j.buildenv.2005.01.007

Yukl, G. (1971). Toward a behavioral theory of leadership. *Organizational Behavior and Human Performance,* 6(4), pp. 414–440.

Yukl, G. (1989). *Leadership in organisations.* Englewood Cliffs, NJ: Prentice-Hall.

Zakaria, I. B., Mohamed, M. R. B., Ahzahar, N. & Hashim, S. Z. (2015). A study on leadership skills of project manager for a successful construction project. *International Academic Research Journal of Social Science,* 1(2), pp. 89–94.

Part IV

Country perspective on monitoring and evaluation practices

9 Monitoring and evaluation in developed countries
A global view

9.1 Abstract

The aim of this chapter is to provide an understanding of the monitoring and evaluation (M&E) practices of the United Kingdom (UK) and Australia, referred to as developed economies using literature review approach. The contribution of M&E in developed economies, particularly the UK and Australia, shows strong linkage towards project performance improvement and accountability in both the public and private sectors of their economies while measures to curtail the limited challenges are seriously implemented. M&E in the UK and Australia have received tremendous acceptance and improvement across many economic sectors especially in construction and health. Lessons learnt from developed economies shows well-established systems and frameworks such as the logframe are utilized in the M&E implementation process to engender effective M&E communication among the project team and policy implementers; allocation of adequate resources for M&E has necessitated for the routine M&E of projects and policies by effective institutional leadership in the M&E process; and the gradual incorporation of M&E in all departments has resulted in the improved benefit on project and programmes. These lessons will serve as good grounds for continuous learning and improvement to shape the M&E practice in developing economies.

9.2 Introduction

This section of the book provides a global view of the M&E practice. The Australian and UK perspectives are reviewed. These countries are categorized as developed economies based on the United Nations' classification (World Economic Situation and Prospects, 2014). In developed economies such as those of the UK and Australia, M&E is viewed as a key element in the transformation of the economy, ensuring that the public sector is effective, efficient and responsive to the citizenry and parliament. A system such as the logical framework (logframe) is one of the most common approaches used in developed countries for project management in both the planning and monitoring of projects. The logframe matrix is a tool that is applicable to all organizations, both governmental and non-governmental, that are engaged in development activities (Martinez,

2011). Hummelbrunner (2010) further opines that regardless of the many criticisms of the logframe, it has remained a widely used planning and monitoring tool owing to its outstanding contribution to project implementation by many project donors. Myrick (2013) however expresses the view that a pragmatic approach to M&E is ideal in the real world: practitioners may be limited by constraints that will prevent their continued use of either a logframe or some overly pragmatic approach to M&E. He further explains that whatever the approach is used, at least the basic principles for M&E, namely that it is measurable, objective, a performance indicator and with target and periodic reporting, should be ensured.

The advantages of a logframe include simplicity and efficiency in data collection, recording and reporting (Kamau & Mohamed 2015). Other approaches include stochastic methods, the fuzzy logic model and miscellaneous methods (Kamau & Mohamed, 2015). Of all the methods, the earned value analysis (EVA) has remarkable advantages in accuracy, flexibility and adaptability for project complexity. This may have contributed to the Malaysian government's deciding to implement EVA to enhance the level of project management for the whole country (Abdul-Rahman, Wang & Muhammad, 2011). An M&E budget needs to be developed and included in the overall project budget to provide the M&E function with its due recognition in its place in project management (Gyorkos, 2003). This is the case in developed countries such as France, Germany and the majority of Europe. Apart from the framework provided, politics is also a major element to take into consideration in projects. Rogers (2008) advocates for multi-stakeholders' dialogues in the data collection, hypothesis testing as well as an intervention to secure greater participation. Monitoring is integral to project management, hence very complex, resulting in confusion in its implementation on projects (Crawford & Bryce 2003). Monitoring as such enhances the project management decision-making during the implementation phase, thus securing the success of the project (Crawford, & Bryce, 2003; Gyorkos, 2003).

In any project, the progress of work is required to be monitored and compared as the work proceeds to be able to identify and measure differences in the originally planned objectives. This is the case in developed countries. There have been significant improvements, according to Cracknell (1994), in the M&E during the last decade of public-sector investment in the UK as well as Australia. This he attributes to the great importance attached to M&E at their prime ministerial level in the 1980s. It is difficult and most times impossible to manage project activity or a programme or portfolio effectively when access to accurate and timely information is deficient or even denied. The Treasury of these developed countries performed a key coordinating and motivating role, but not in any way attempting to regiment procedures in a fixed method. It is widely accepted, although there is no proof of the fact that all these assets of effort and resources have resulted in significant improvements in the efficiency and effectiveness of public-sector investment (Cracknell, 1994). There is still scope for further improvements, particularly in the fields of more methodical project-cycle management, better evaluation methods and more efficient reporting of the results from M&E.

According to Cracknell (1994), only a few departments in government set-tings in the world have well-established systems of M&E. Among such depart-ments are the Department of Trade and Industry, the Department of Health, the Overseas Development Administration and the Manpower Services Commission in the UK in a quest to improve the economic and financial assessment of pro-jects and programmes in the country. During the early 1980s in the UK, the Prime Minister, then Mrs. Thatcher, called for the introduction of some form of target-oriented planning in the civil service to motivate their employees and improve their performance. This became popularly recognized as the Financial Management Initiative (FMI). The Treasury then took over the new thrust of the civil service management and the newly created evaluation unit and began to collect material on M&E systems and produced very useful guidelines to be used throughout the civil service. Cracknell (1994) indicated that such initiatives stimulate important developments in M&E systems in the public sector.

In the Australian public service, M&E provides a sound base for assisting the bureaucracies of developing nations and helping them build their capacities to an improved level of governance in general. This is based on a good and well-en-trenched evaluation and accountability culture. Although the country is not at the forefront of M&E systems' capacity building, substantial expertise is readily available, particularly in the consulting industry (AusAID, 1997).

9.3 Overview of the United Kingdom construction industry

The UK construction industry is an internationally recognized industry as one of the largest in Europe and serves as a world-class benchmark for most developing and other developed economies (Department for Business Innovation & Skills, 2013). The industry further contributes nearly £90 billion (6.7%) in added value to the UK economy, covers over 280,000 businesses, comprising some 2.93 mil-lion jobs which translate to about 10% of the total UK employment (Department for Business Innovation & Skills, 2013). The UK's construction industry key per-formance indicators (KPIs) ensured that industry performance, client satisfaction, quality, cost and health and safety indicators are achieved. This notwithstand-ing, the Construction Products Association (CPA) (2017) informs of the current down surge and the slow growth of the industry, reaching its lowest in the last six years. This the CPA attributes to the run-up to the UK's exit from the European Union (EU) (CPA, 2017). The UK construction sector is made up of three main sub-sectors, namely the construction contracting industry, the provision of con-struction-related professional services and construction-related products and materials. The nature of activities, their size relative to the gross value added (GVA) and level of employment are presented in Figure 9.1.

9.4 Overview of the Australian construction industry

Similar to the UK construction sector, Australia's construction industry is described as a major driver of economic activities (AI Group, 2015). It is also the

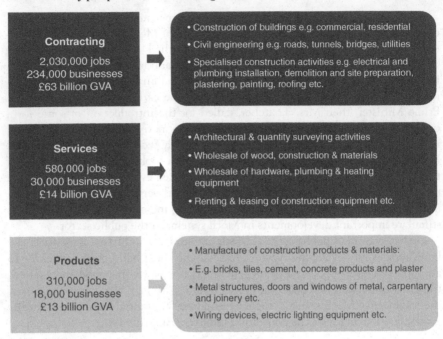

Figure 9.1 Composition of the UK construction sector.

Source: Department for Business Innovation & Skills, 2013

third-largest industry after mining and finance sector and contributes about 8% of the GDP in value-added terms. It comprises over 330,000 businesses nationwide and directly employs over one million people, which represents about 9% of the total workforce. It produces the buildings and infrastructure that are indispensable to the survival of all other industries (AI Group, 2015). The Australian construction industry affects many economic sectors such as manufacturing (materials, equipment components), services (engineering, design, surveying, consulting, lease management) and traditional construction trades. Figure 9.2 presents the industry's output, size and share of the GDP.

9.5 Philosophical basis and policy for monitoring and evaluation in the United Kingdom and Australia

The fundamental philosophy that formed the basis of the introduction of target-oriented planning in the civil service in the UK contributed enormously in the policy of privatizing parts of the public sphere which could be handled as well by the private sector (though on the policy grounds that the restraint of having to make a profit provides an automatic target and measure of accomplishment). This underlying principle is less noticeable in the case of the naturally monopolized sector to private establishments such as the privatization of the British

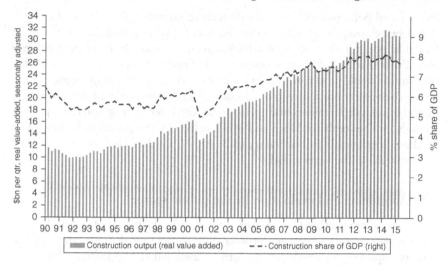

Figure 9.2 Australian construction industry output, size and share of GDP.

Source: AI Group, 2015

Gas (OFGAS) and the British Telecommunications (OFTEL), but in these cases government also set up publicly regulatory bodies ensuring that there is no abuse of the monopoly power given to the companies.

Towards the end of the 1980s and in the early 1990s, privatization policy became an issue of great concern and so was vigorously pursued. This led to initiatives such as "Next Steps". This fostered the setting up of independent or semi-independent agencies that took over blocks of work that had till then been handled by civil service departments. The "Market Testing" was another initiative which tested the tasks of the current performance of civil servants to ascertain whether they might not be done more cost-effectively by the private sector. These policies were the sole responsibility of the efficiency unit (cabinet office). The outcome of this move to privatizing M&E functions was two-fold. First, it enhanced the necessity for M&E in order to accurately account for all public funds and of the necessity also to prove that government investment is worth competition with the private sector counterparty. Secondly, it may make it more difficult for the main department to ensure that adequate M&E is carried out by the required agencies. This is simply because control of authority is not direct as it ought to be, especially in the case of privatized agencies where there is virtually no control at all.

Contemporary approaches were worked out for these M&E activities controlled by the agencies or mainly by private organizations. If the agency could operate as a profit-making organization, the necessity to establish targets would be reduced, but if not (which was the frequent case), it is paramount that the agencies would be given clear targets. This may be difficult most times because in the public sphere, there are frequently several objectives and most times, these objectives are of social policies which are very difficult to quantify. The obscured factor is that agencies frequently have to exercise a coercive responsibility in the quest for public policy and to make sensitive judgments that could affect the welfare of individuals involved. Selecting operationally feasible recital targets in such circumstances would most times be difficult, but very vital if the main department is to exercise an accountability role adequately. Renegotiations of performance contracts between departments and agencies are also essential because these provide grounds for the results of M&E and M&E is likely to be particularly influential. Further experience is still necessary to be gained for the M&E exercise to adapt to its systems in the years ahead.

9.6 Monitoring and evaluation in the public sector

As long as a monitoring process/exercise is a major factor for effective delivery, it is paramount that various government departments and units carry out this activity, especially in developed countries. Cracknell (1994) informs that M&E is a natural function of management. At the operational level of monitoring, data or information are urgently provided for the evaluation process. Evaluation is seen as a vital management procedure, hence the significant difference between both activities. While monitoring is general, evaluation is essentially selective (Cracknell, 1994). Resources available to agencies mandated to undertake M&E most often allow for relevant and specified evaluations to be done. Earlier, it was usual to evaluate "once-in-a-lifetime" projects, but currently, there are emergent tendencies to routinely evaluate projects, programmes and even portfolios, which could come up with useful lessons for future reference. The evaluation function is the responsibility of a particular unit or section in the public organization and it has to develop or start in those departments that spend considerable amounts of public funds supervised by effective M&E leadership. This is vital because of the necessity for accountability and also because these government departments are heavily into project assistance and needed a feedback mechanism as to the effectiveness of these project appraisal techniques in use, giving impetus to the effective communication in the monitoring and evaluation process. On the other hand, its development is at a slower rate when compared to those in ministries that spend public funds directly, usually in the form of nationwide programmes/portfolios rather than easily identified projects. In this case, accountability appears to be less and M&E's main purpose which is to enable improved performance in implementing broad policies is defined.

In recent years, the pressure for the privatization of these departments has provoked these ministries to demonstrate their cost-effective skills at a reasonably high level. This enhanced the status and importance of M&E in both the public and private sectors. Under these circumstances, evaluation is bedeviled by the problems of coping with competing (and sometimes incompatible) objectives, which are often rather abstract, such as those of welfare, security or health. M&E in the public sector in the UK, in particular, has made tremendous advances in recent years, with no sign that the momentum has slackened (Cracknell, 1994), though many challenges are still noticed in the continued development of evaluation techniques. One of these challenges is the self-view of some policy-makers in the government that policies on M&E may result in a complicated trade-off between competing interests, which cannot be evaluated in any scientific way. While Cracknell (1994) shares this view, he argues that there is plenty of scope for evaluating the effects of policy, even while shrinking from evaluating policy decisions directly and the lessons so learned can be a valuable input into future policy decisions.

Another major challenge among the potential political leaders is the fear that evaluation policy may reduce their selfish interests and may to a large extent make it difficult for future trade-offs that often typify policy-making. It is by no means uncommon for ministers to indicate that they do not wish particular policies to be evaluated. Here again, the best response may be simply to stress the value of learning what has been the outcome of past policy decisions as one of the inputs that go into the making of fresh policy decisions. None of these obstacles seems likely to prevent the continued development of public-sector M&E in the UK. Recent evidence is that M&E is becoming fully incorporated into the existing administrative and management procedures. Evaluation expertise is on an increase with well-published evaluation reports that provide a good basis for better informed public debate. Cracknell (1994) posits that M&E is well past its probationary period and has now established itself as a vital part of public sector investment management in the UK and Australia.

9.7 Monitoring and evaluation policy challenges in developed countries

Regarding the rapid acceptance of M&E systems, it is not without doubt that many are having difficulty in living up to the ambitious demands placed on them. Many projects or programme M&E systems have been criticized for their inadequacy and narrow efficiency. For instance, information arrives rather too late in most cases, sometimes the information does not answer the right questions and sometimes the information is costly to retrieve. This therefore requires a quick response by deploying strategies for effective communication of M&E data and findings. In other cases, the attention is narrowly focused on certain quantitative and financial aspects of the projects and most of the information refers only to the period of physical implementation. There are other challenges such as overly focusing on monitoring of project implementation; limited studies on how programmes operate, how they are sustained or whether they can produce intended impacts; capital budgeting being the focus rather than recurrent budgeting; M&E units being located in agencies created to oversee implementation; and short-term planning and budgetary cycles leading to a focus on short-term implementation objectives. The capacity of M&E leadership is noticed as a major pointer to the overall M&E process. Effective M&E leadership will drive the effective planning and implementation of M&E for the desired outcomes.

9.8 Findings and lessons learnt

The contribution of the implementation of M&E in the UK and Australia as per the review above shows strong linkage towards project/programme performance improvement and accountability in both the public and private sectors of their economies while measures to curtail the limited challenges are seriously implemented. M&E in the UK and Australia have received tremendous acceptance and improvement across many economic sectors such as construction and health. Therefore, based on the above review, the following lessons can be deduced:

1   Well-established systems and frameworks such as the logframe are utilized in the M&E implementation process to engender effective M&E communication among the project team and policy implementers;
2   Allocation of adequate resources for M&E has necessitated for the routine M&E of projects and policies by effective institutional leadership in the M&E process;
3   The gradual incorporation of M&E in all departments has resulted in the improved benefit on projects and programmes.

Summary

In this chapter, a review of monitoring and evaluation practice in a developed country context, namely the United Kingdom (UK) and Australia, was presented. The chapter revealed the nexus of effective communication and leadership in delivering effective monitoring and evaluation in the context of the UK and Australia. The various tools for effective M&E were reviewed. Also, an overview of both countries' construction industry/sectors was presented. Further, a discussion on the philosophical basis and policy regulation guiding M&E in both countries was presented. Finally, the challenges, findings and lessons learnt were all captured in this chapter. The next chapter focuses on the review of M&E practice in developing countries, namely Kenya and South Africa.

References

Abdul-Rahman, H., Wang, C. & Muhammad, N. B. (2011). Project performance monitoring methods used in Malaysia and perspectives of introducing EVA as a standard approach. *Journal of Civil Engineering and Management*, 17(3), pp. 445–455, doi:10.3846/13923730.2011.598331

AI Group (2015). *Australia's construction industry: Profile and outlook*. Economics Research.

Cracknell, B. E. (1994). Monitoring and evaluation of public-sector investment in the UK. *Project Appraisal*, 9(4), pp. 222–230, doi:10.1080/02688867.1994.9726955

Crawford, P. & Bryce, P. (2003). Project monitoring and evaluation: A method for enhancing the efficiency and effectiveness of aid project implementation. *International Journal of Project Management*, 21(5), pp. 363–373, doi:10.1016/S0263-7863(02)00060-1

Department for Business Innovation & Skills (2013). *UKconstruction: An economic analysis of the sector*. Available online at www.bis.gov.uk. Retrieved on 31st August 2018,

Gyorkos, T. W. (2003). Monitoring and evaluation of large scale helminth control programmes. *Acta Tropica*, 86(2–3), pp. 275–282, doi:10.1016/S0001-706X(03)00048-2.

Hummelbrunner, R. (2010). Beyond logframe: Critique, variations and alternatives. In: Fujita, N. (Eed). *Beyond Logframe: Using Systems Concepts in Evaluation*. Presented at the Issues and Prospects of Evaluations for International Development, pp. 1–34. Japan: Foundation for Advanced Studies on International Development.

Kamau, C. G. & Mohamed, H. B. (2015). Efficacy of monitoring and evaluation function in achieving project success in Kenya: A conceptual framework. *Science Journal of Business and Management*, 3(3), p. 82, doi:10.11648/j.sjbm.20150303.14

Martinez, D. E. (2011). *The logical framework approach in non-governmental organizations*. CA: University of Alberta.

Myrick, D. (2013). A logical framework for monitoring and evaluation: A pragmatic approach to M&E. *Mediterranean Journal of Social Sciences*, doi:10.5901/mjss.2013.v4n14p423

Rogers, P. (2008). Using programme theory to evaluate complicated and complex aspects of interventions. *Evaluation*, 14(1), pp. 29–48, doi:10.1177/1356389007084674

World Economic Situation and Prospects (2014). *Country classification: Data sources, country classifications and aggregation methodology.*

10 Monitoring and evaluation in developing countries

An African experience

10.1 Abstract

This chapter reviews the literature on the monitoring and evaluation practices of two developing economies, i.e. Kenya and South Africa. The literature informs that while the field of M&E practice continues to grow and brings professionals together under specialized departments to operate, the institutional framework for M&E practice remains weak. Also, in an attempt to safeguard the foreign resource interest of the developed world and the emphasis on output, outcome, accountability and transparency of investments contributed to the rise in the demand for M&E in Africa. Lessons learnt from the review indicate that Kenya and South Africa have a well-structured M&E system, plans and processes for national, provincial and municipal infrastructure delivery with clearly stated M&E responsibilities. The regulatory, policy environment and support for M&E appear sufficient to give independence to institutions to effectively monitor and evaluate performance. Also, the South African Construction Industry Development Board (SACIDB) is well integrated with the national government-wide M&E structure and with the support of other constitutional bodies, exercises oversight responsibility to ensure the growth, development and sustenance of the South African construction industry. These are relevant for adoption and implementation by other developing nations.

10.2 Introduction

This section of the study presents an understanding of M&E in developing countries with a focus on the Republic of **Kenya** and **South Africa**. The countries were selected based on the United Nations' classification of economies with low per capita levels and which are less industrialized (World Economic Situation and Prospects, 2014). The purpose of this review is to understand the M&E practices in the context of other developing nations for lessons to be learnt and for the transfer of knowledge for the benefit of other cultural contexts.

10.3 Monitoring and evaluation in Africa

M&E in Africa dates back to as early as the 1990s and has seen a steady and fast growth on the continent (Basheka & Byamugisha, 2015). M&E has also evolved over the years as a field of practice, profession and academics. Basheka and Byamugisha (2015) aver that while the field of practice continues to grow and brings professionals together under specialized departments to operate, the institutional framework for M&E practice remains weak. Also, more than thirty national evaluation associations exist globally. However, before the African Evaluation Association (AfrEA) was established in 1999, the Ghana M&E Forum (GMEF) had long since been established in 1997, thereby becoming the first African national evaluation association to be established. It is believed that several factors contributed to the development of the AfrEA.

The rather fast growth of the evaluation association on the continent explains the need for and importance of evaluation (Mertens & Russon, 2000). Nonetheless, the influence of the developed world, particularly America, on Africa to adopt the M&E practice, particularly in governance, could only be an attempt to safeguard the foreign resources channelled into the governance of developing countries. This growth is evident in the over 500% increase in the numbers of national evaluation associations, with much of this growth occurring in developing countries between the years 1995 and 2000 (Basheka & Byamugisha, 2015). Also, due to the political recognition of the utilization of evaluation in governance and the emphasis on output and outcome, accountability and transparency, there was an increase in demand for M&E in Africa (Basheka & Byamugisha, 2015).

Evaluation is undertaken in all fields in Africa, particularly in all sectors of governance. Most government ministries and departments in Africa have established M&E units responsible for the M&E of policies, programmes and projects to ensure efficiency and accountability. The construction industry in Africa attracts heavy funding from governments and hence the need for M&E for efficiency and accountability of limited resources.

10.4 Kenya

Kenya is an East African country and shares borders with Ethiopia to the north, Uganda to the west and Sudan to the north-west. To the east, Kenya shares a border with Somalia and the Indian Ocean towards the southeast (see Map 10.1). The Republic of Kenya is made up of 47 counties for devolution of governance and development and to ensure that development is sent to the doorstep of its citizens. Kenya is usually referred to as the economic hub as it serves as a host to some international corporations' African headquarters and booming construction industry (Wanjira, 2016).

10.4.1 Construction industry outlook of Kenya

It is recorded by the International Construction Market Survey report that infrastructure grew by 13.6% in the year 2015 (McGuckin, 2017). This significant growth was attributed to the mega-infrastructure and energy flagship projects initiated by the government as part of robust infrastructure development of the country. This growth is no doubt a major contributor to the larger economy

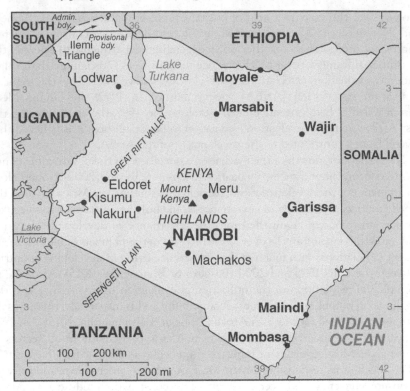

Map 10.1   Map of the Republic of Kenya.

(Wanjira, 2016). The Kenyan construction industry (KCI) is a major economic driver along with manufacturing and information communication technology which was responsible for an improved GDP) growth of 6.0% in 2016, compared to 5.6% in the previous year (Odero, Reeves & Chokerah, 2017).

Owing to the reported vast deficit in major infrastructure development in Kenya such as rail, roads and ports and the rapid growth in population degenerating into rising demand for housing in most parts of the country, there is a significant cause for the continuing growth of the building and construction sector (Wanjira, 2016). This notwithstanding, just as in other developing countries, similar characteristics exist in the challenges facing the Kenyan construction industry (Ofori, 2000). Major challenges include capital sourcing for infrastructure development, quality assurance challenges, corruption in the selection of contractors, unskilled labour and occupational, health and safety challenges (Wanjira, 2016). Further, Auma (2014) identified poor quality, late completion and cost overruns as performance challenges confronting construction project delivery in Kenya. It is therefore vital for the Kenyan construction industry to have a measure with which to assess the performance of the industry.

10.4.2 Monitoring and evaluation regulatory policy in Kenya

The Kenyan Constitution fundamentally requires compliance with principles of good governance and transparency in the conduct and management of public programmes and projects (Ministry of Devolution and Planning, 2016). It is, however, reported by Mulama, Liguyani and Musiega (2014) that there was project-based M&E as early as the 1980s until the inception of a government-wide M&E system in Kenya. It is further argued that early government-wide M&E attempts were made by the International Monetary Fund (IMF) and the World Bank but these were ineffectively implemented in 2000 in the Interim Poverty Reduction Strategy Paper (I-PRSP) (2016–2018) (Mulama et al., 2014; Mugo & Oleche, 2015). Hence, through the government-wide M&E, Kenya is currently implementing two M&E systems. They are the National Integrated Monitoring and Evaluation System (NIMES) and the County Integrated Monitoring and Evaluation System (CIMES) which monitors all government programmes and the Kenyan Vision 2030 via its Medium-Term Plans (MTP) as well as the County Integrated Development Plans (CIDPs) which will provide the government with a reliable policy implementation response to efficiently apportion resources over time. An M&E framework is designed to meet the requirements of an M&E policy of the Kenya government to guide the implementation of the National Integrated Monitoring and Evaluation System (MPND, 2012).

The coordination responsibility of implementation lies with the Monitoring and Evaluation Directorate (MED) of the Ministry of Planning, which aims to improve management for development results. The Kenyan MED was established to provide a three-tier relationship between the MED at the national level, the Central Project Planning and Monitoring Units in each line ministry and the sub-county planning units created in each district. The MED also offers leadership and manages the system by ensuring that the Annual Public Expenditure Review (PER) and the Annual Progress Reports (APRs) on the Medium-Term Plan of Vision 2030 are made available timeously (Mulama et al., 2014). The M&E policy highlights the need for results, accountability, efficiency and transparency as vital ideologies towards public programmes and projects management in Kenya. The purpose is to enable reporting and feedback on the implementation of development programmes and projects at the sub-county, county and national levels. Programme and project implementers under the policy will diligently collect and compare qualitative and quantitative data of implementation progress against the set goals, to ascertain the progress made towards meeting set objectives. The process of monitoring will be guided by indicator identification, indicator data, frequency of data collection, responsibility for data collection, data analysis and use and reporting and dissemination.

At the county and national levels, monitoring will focus on assessing progress made towards achieving the sectorial development outcomes. On the other hand, the evaluation will answer specific questions about the performance of development interventions. It will further focus on the reasons for the results' achievement or otherwise. The policy requires that external evaluations will be conducted by an accredited and registered independent body and stakeholder in a participatory process, whereas internal evaluations will be conducted by the implementing agency using rapid appraisal methods. Effective M&E are founded on a clear, logical pathway of results in which results at one level lead to results at the next level. Results from one level flow towards the next level, resulting in the achievement of the overall goal. If there are gaps in the logic, the pathway will not flow towards the required results. The major levels that the policy focuses on are inputs, outputs, including processes, outcomes and impacts.

10.4.3 Construction project monitoring and evaluation in Kenya

M&E practice is widely acknowledged in the Kenyan construction industry as a critical determinant for successful project delivery and has shown a positive impact on the successful implementation of projects when undertaken efficiently. The literature revealed that road construction projects in the Kenya National Highways Authority are influenced positively by M&E (Gitahi, 2015). Also, the impact of M&E factors is seen in the success of development projects (Kamau & Mohamed, 2015; Wachaiyu, 2016). In a related study, Mwangu and Iravo (2015) studied how M&E affect the outcome of constituency development fund projects in Kenya while Kiarie and Wanyoike (2016) revealed a positive correlation of M&E on the successful implementation of government-funded projects in Kenya. It is therefore suggested that the effective implementation of M&E will result in successful project delivery in the Kenyan development process (Mwangu & Iravo, 2015).

M&E has been understood as a process, system or an approach and for it to be effective and efficient, relevant factors must be present to ensure the process or system generates the level of success set for the project. According to Ogolla and Moronge (2016), sufficient budgetary allocation and key stakeholder involvement in M&E are imperative to ensure that government water projects in Kenya are successful. Also, in the implementation of constituency development fund projects in Kenya, Mwangi et al. (2015) placed emphasis on the need for enough budgetary allocation and the involvement of key stakeholders. They further revealed a 28% increase in M&E effectiveness with a one-unit increase in the technical competence of M&E staff. Onjure and Wanyoike (2016), on the other hand, revealed M&E practices such as the quality of data collected and analyzed, the appropriate use of M&E tools and the project team efforts influenced the performance of national government-funded construction projects. Another important determinant of M&E is communication (Mugambi & Kanda, 2013) while the leadership of the project manager is reported to affect the level of effectiveness of the implementation of M&E on projects (Njama, 2015).

10.4.4 Challenges in construction project monitoring and evaluation in Kenya

According to Mulama et al. (2014), a planned M&E system structure for project implementation existed for most of the social development projects with collaborative funding between government and international development agencies and executed through the respective line ministries. However, these projects take a long time to reach completion as well as there being the problem of overspending beyond project budget (Mulama et al., 2014). Ogolla and Moronge (2016) argue that a significant challenge for the poor performance of the water development project in Nairobi County, regardless of the huge investment by the government, is as a result of ineffective M&E processes. Mugo and Oleche (2015) also posit that if the right resources for M&E such as the release of funding and skilled staff are not present at the right time, the M&E system serves no purpose as data needs to be collected at such times that will inform effective evaluation and decisions to be taken on the project. They further stressed that the delay and inaccurate information generated by the ineffective M&E and failure to adopt M&E requirements were major challenges to M&E (Mugo & Oleche, 2015).

Likewise, Wanjira (2016) mentions the shortage of relevant skilled construction professionals who possess the required skills for the M&E of project implementation due to poor remuneration. Also, the time allocated for M&E is not defined, making M&E an ad-hoc practice in the Kenyan construction industry. Further, he argues that the unstructured project implementation organizations make communication and coordination of the M&E ineffective (Wanjira, 2016).

10.4.5 Findings and lessons learnt from Kenya

From the foregoing, M&E of projects in Kenya has revealed the integration of an M&E system at all levels of the project implementation structure. This encompasses the national level (NIMES) and county and district levels of project implementation which are coordinated by the MED. The literature reviewed in the Kenyan context also revealed that M&E practice contributes to the achievement of successful project implementation. Factors that influence the effectiveness of M&E practice in the Kenyan building and construction industry include but are not limited to adequate budgetary allocation, stakeholder involvement, technical competence and the appropriateness of M&E practices. A major challenge, however, to the implementation of M&E in the Kenyan building and construction industry is the ineffectiveness of the M&E system and processes which generate late and inaccurate M&E information for management decision-making. Also, the late release of budgetary allocation and the failure to adapt to M&E requirements have rendered the M&E of projects unacceptable; hence the need for strategies to strengthen M&E systems and guidelines for construction project M&E to ensure project success.

10.5 South Africa

Map 10.2 shows South Africa, a nation located at the southernmost end of the continent of Africa. South Africa is bordered by countries like Namibia, Botswana, Zimbabwe and Mozambique all in the north. Lesotho and Swaziland are sovereign nations located in South Africa. The South Atlantic Ocean and the Indian Ocean borders South Africa in the south. The country has an overall

Map 10.2   *Map of the Republic of South Africa.*

coverage of 1,219,090 square kilometers which comprises 1,214,470 square kilometers of land and 4,620 square kilometers of water (The CIA World Factbook, 2016). The middle-income country according to the World Bank has the biggest stock exchange market in Africa and among the top twenty countries in the world. The country is also blessed with abundant natural resources, well-developed financial, legal, communications, energy and transport sectors.

10.5.1 Construction industry outlook of South Africa

The impact of the South African construction industry on the economy cannot be overstressed. Indeed, the South African Construction (2016) acknowledges the significant contribution of the construction industry to employment and growth in South Africa. Pillay and Mafini (2017) aver that the construction industry is the major contributor to the South African economy. Statistics South Africa (2016) cited by Pillay and Mafini (2017) indicates that the construction industry contributes significantly to job creation for the majority of the working population in the country and also to the GDP of the country. The Construction Industry Development Board (CIDB) (2016) reports that nearly 50% of the national capital and about 4% of the nation's GDP is as a result of the contribution of the construction industry. It is, however, acknowledged that the key contributor to the expansion of the construction industry is the expenditure of government to the sector. This is evident in the government announcement in the 2015 budget statement to invest R813.1bn on major infrastructure projects in the transport, water and energy sectors (Nene, 2015). Also, the Human Settlements Vision 2030 and the 2023 mud-school building replacement project will ensure potential government capital injection into the construction industry (Report Buyer, 2016).

This notwithstanding, a depressed construction industry outlook (SA Construction, 2013; 2014; 2015; 2016), Pillay & Mafini (2017) identifies a supply chain management challenge in the South Africa construction industry. They further outline the challenges as skills and qualifications in the industry, procurement practices and systems, supply chain integration, supply chain relationships and the structure of the construction industry. Windapo and Cattell (2013) also identified twelve challenges influencing the performance, growth, and development of the South African construction industry through an extensive literature search. The study subsequently through validation established that out of the twelve challenges, increase in the building materials cost, access of affordable mortgage/credit, high interest rates, high rates of enterprise failure/delivery, capacity and performance and the mismatch between available skills and required skills as most significant challenges facing the South African construction industry (Windapo & Cattell, 2013).

In a related study, Windapo (2016) again opined the shortage of skilled or qualified tradesmen such as plumbers, electricians and carpenters in the South African construction industry but was quick to blame the shortage on the lack of high-quality basic education and certification, the compulsory certification of tradesmen and the ageing workforce. Another challenge facing the South African construction industry is the poor quality of work output (Windapo, 2016). According to Bowen, Edwards and Cattell (2012), the perception is that corruption in the South African construction industry is widespread. These corrupt practices common to the industry include conflict of interest, tender rigging, fronting and kickbacks. Khumalo, Mashiane and Roberts (2014) also revealed poor profitability as a major challenge when they studied the harm and overcharge of a precast concrete product cartel in South Africa. The phenomenon was attributed to the excessive cost overruns (30%) and change orders (8.3%) on precast concrete projects in South Africa. The South African industry outlook presented above shows the critical need for the industry to put structures in place to regulate practices in order to ensure that the challenges are mitigated for effective project implementation. The challenges observed in the South African construction industry are similar to those experienced in the Ghanaian construction industry.

10.5.2 *Monitoring and evaluation regulatory policy in South Africa*

Similar to many other developing countries such as Ghana and Kenya, South Africa has an official ministry responsible for M&E and also an M&E unit established in each government department (Abrahams, 2015). The Public Service Commission (PSC) and the Department for Planning, Monitoring and Evaluation (DPME) have been established as statutory organizations with responsibility for M&E the performance and service delivery of government (Dassah & Uken, 2006). The PSC derives its powers from Sections 195 and 196 of the 1996 Constitution of South Africa. As a part of their mandate, the PSC is responsible for investigating, monitoring and evaluating the performance of all programmes of the South African public service. The investigatory M&E role is predominately centered on the financial accountability of the public service.

According to Engela and Ajam (2010), the Government-Wide Monitoring and Evaluation System (GWM&ES) was established in 2005 to address the fragmentation in government M&E. Abrahams (2015) argues the GWM&ES served as a "system of systems" as it extracted germane information from all other independent M&E systems of all government departments. The DPME was also established in 2010 by the Presidency and headed by a minister. The DPME is responsible for the implementation of the National Development Plan (NDP), the Medium-Term Strategic Framework (MTSF) and the five-year implementation plans and adopts a citizens' feedback mechanism as the M&E tool. The prime focus of the DPME is, however, on the implementation of the government policy to gather relevant information for management decision-making to inform future policy direction. The South African Monitoring and Evaluation Association (SAMEA) was also established in 2005 as an umbrella body of individuals and organizations involved in the evaluation functions. SAMEA provides support and guidance and strengthens the development of M&E as a profession, as an industry and also as an independent voice (governance tool) providing expert advice to the DPME (Abrahams, 2015; Basson, 2013).

In summary, legislative policies and an enabling environment for M&E in the South African public sector include but are not limited to the following:

   i The Constitution of the Republic of South Africa, 1996
  ii Batho Pele White Paper, 1997
 iii Public Finance Management Act (Act 1 of 1999)
  iv Treasury Regulations, 2002
   v Policy Framework for a Government-wide Monitoring and Evaluation System, 2007
  vi National Treasury Framework for Managing Programme Performance Information, 2007
 vii South African Statistical Quality Assessment Framework (SASQAF), First edition (2008) and Second edition (2010)
viii Green Paper on National Performance, 2009
  ix Guide to the Outcomes Approach, 2010
   x National Evaluation Policy Framework (NEPF), 2011; and
  xi Performance Monitoring and Evaluation: Principles and Approach, 2014

(Mtshali, 2015).

10.5.3 Municipal infrastructure project implementation and M&E in South Africa

In South Africa, municipal infrastructure project delivery is guided by a legislative, policy and institutional framework. The framework outlines a structured process in infrastructure delivery at the local level, provincial level and the national level. While projects are implemented at the local (municipal) level, there is the need for supporting roles at the provincial and national levels since funding for infrastructure delivery is made available through the central government. The responsibility of municipal infrastructure project delivery is distributed to the national departments. These responsibilities suggest that the Department of Provincial and Local Government (DPLG) monitors the cross-cutting conditions and overall progress of projects executed while the respective sectorial departments such as the Department of Transport (DoT) and the Ministry of Public Works monitor the performance of municipalities regarding sector-specific criteria and the overall sustainability of sector infrastructure.

Also, the Department of Public Works is tasked with the monitoring of the poverty alleviation criteria whereas the responsibility for the financial reporting and revenue-related criteria lies with the National Treasury (Department of Provincial and Local Government, 2006). At the local level, three distinct levels are established with each level performing a specific task in the infrastructure delivery process. These are the project level, the local sector level and the municipal level. The provincial level has two separate levels, namely the provincial sphere and the provincial sectors, while at the national level, the national sphere and the national sector describe the two distinct levels for implementing project task in the delivery process. Figure 10.1 shows the four-phase infrastructure project (service) delivery life cycle, namely the policy, planning, implementation and service provision phases. The implementation phase describes the M&E role of the local, provincial and national stages.

10.5.4 The South African Construction Industry Development Board

The South African Construction Industry Development Board (CIDB) was established by a Parliamentary Act 38 in 2000. The CIDB was established to provide leadership for the reconstruction, growth and development of the South African construction industry. The Constitution mandates the board to create a nation-wide record of contractors and projects to thoroughly regulate, monitor and improve the performance of the construction industry to ensure sustainable growth, delivery and empowerment; encourage management capacity improvement and the consistent application of procurement policies at all levels of government; help improved performance and best practice of both public and private sector clients, contractors and other stakeholders in the construction delivery process; promote sustainable involvement of the emerging sector; provide strategic direction; and develop effective partnerships for the growth, reform and improvement of the construction sector.

Phase	National Level		Provincial Level		Local Level		
	National Sphere	National Sectors	Provincial Sphere	Provincial Sectors	Municipal Level	Local Sector Level	Project Level
Phase 1: Policy	Develop municipal infrastructure policy and set standards for delivery systems	Develop sector policies, systems, procedures			Service provision policies and bylaws	Sector policies for free basic services	
Phase 2: Planning	Develop framework for National Spatial Development Perspective (NSDP)	Macro sector planning	Provincial Growth and Development Strategies (PGDS)	Provincial Sector Plans	IDP	Local sector plans	Project Pre-Feasibility and Feasibility Studies and Business plans
Phase 3: Implementation	Municipal infrastructure programme management, collaboration, mobilise support and monitoring	Monitor implementation of systems, procedures, and collaboration around support	Monitor implementation of infrastructure policy and delivery systems and mobilise and co-ordinate support	Monitor implementation of systems, procedures, and collaboration around support	Infrastructure delivery systems put in place and project management	Technical department (eg. water, roads etc. oversee project implementation)	Project cycle – implement technical systems, procedures
Phase 4: Service Provision	Regulate and oversee systems and procedures	Regulate and oversee sectoral systems, procedures	Systems and procedures support	Service provision support and intervention	Regulate and oversee sectoral systems, procedures	Regulate and oversee sectoral systems, procedures	Service provision (O&M)

Figure 10.1 A comprehensive infrastructure delivery implementation framework.

Source: Department of Provincial and Local Government, 2006

In line with the mandate of the CIDB and with the support of other organizations such as the South African Public Works Department (PWD) and the South African Council for Scientific and Industrial Research (CSIR) performance indicators are formulated which are linked to the best practice standards and guidelines and which serve as the basis for measuring the performance of the South African construction industry at separate project levels (Marx, 2014). The Construction Industry Indicators (CII) established by the CIDB and the PWD serve as the standards for the performance of construction firms and all other related stakeholders for the M&E of the performance of their function in the industry. Also, in the M&E function of the CIDB to monitor and evaluate industry performance and support contractor development, the CIDB publishes on a regular basis the CIDB Compliance Monitor, CIDB Quarterly Monitoring (now Construction Monitor), CIDB Business Condition Survey and sector-specific status reports (CIDB, 2014).

According to the CIDB (2017a), the Construction Monitor offers a summary of the South African contracting sector with the emphasis on supply and demand, contractor development, employment and transformation. Also, the Compliance Monitor of the CIDB outlines indicators to measure the level of conformity of i-tender/record of projects by public sector client bodies. Clients use indicators to verify their conformity level with CIDB regulations in line with the CIDB programme of action to support compliance and enforcement (CIDB, 2017b). The roles of the CIDB in the South African construction industry generate some reform recommendations to be enacted and implemented for the ultimate growth and development of stakeholders and the industry at large. For example, in the 2015/2016 annual report of the CIDB, the draft report on prompt payment details is published of the critical issues of late payment to construction industry suppliers, which threatens the survival of businesses, especially emerging businesses (CIDB, 2016b). The outlook of the CIDB has been positive in driving the compliance by contractors and stakeholders to the CIIs to maintain some significant level of performance of the industry. Nonetheless, some challenges are faced in the performance of this mandate. These M&E implementation challenges faced by the CIDB, the PWD and other statutory bodies and organizations are considered under the challenges in the construction project M&E in South Africa.

10.5.5 Challenges in construction project monitoring and evaluation in South Africa

Challenges seem inevitable in the M&E of projects, programmes and policies for many reasons. The relationship between M&E and the entire project life cycle is very much integrated. Hence, any challenge faced in the project cycle affects the implementation of M&E. Hence, owing to the similarities of the South African construction industry to the Ghanaian construction industry and to many other developing countries regarding the challenges facing the industry, it is safe to admit that project features such as the uniqueness and complexity of projects, limited and scanty information on projects, poor planning and budgeting and the capacity of constructors pose as great challenges to the implementation of M&E (Tengan & Aigbavboa, 2016).

In 2014, the DPME reported on the empirical findings of a survey amongst the national and provincial government departments regarding the challenges in M&E. Accordingly, the report revealed that there exists no formal culture of M&E amongst government departments as M&E is viewed as a regulating and controlling function rather than a tool for continuous improvement (Republic of South Africa, 2014). Also, it was evident that about 81% of the departments surveyed do not plan or undertake any evaluation of activities. This therefore has delinked the function of planning to inform policy decision making and budgeting at the national and provincial departments. Again, the general understanding of M&E is low amongst government departments and has led to less influence of M&E on the departments' policy priorities (Republic of South Africa, 2014).

Similarly, the survey report by the DPME revealed that the M&E of activities and outputs received much attention, being the focus of M&E while the M&E of outcome and impact of implemented activities were of less importance. Further, poor data quality, low adoption of information technology systems to support M&E and the inadequate capacity of departments to undertake M&E were rated high on the list of challenges. Finally, departments were worried about the lack of a common guiding framework to complement M&E concepts and practices across the South African public sector and the lack of support from the National Policy Framework to enable departments' effective use of the M&E frameworks (Republic of South Africa, 2014).

10.5.6 Findings and lessons learnt from South Africa

In the case of South Africa, M&E of projects have been faced with challenges in its implementation across all the three spheres (national, provincial and municipal). However, some lessons can be learnt:

i   There appear to be a well-structured M&E system, plans and processes for national, provincial and municipal infrastructure delivery with clearly stated M&E responsibilities (Figure 10.1).

ii The regulatory, policy environment and support for M&E appear sufficient to give independence to institutions to effectively monitor and evaluate for performance.

iii The South African Construction Industry Development Board (SACIDB) is well integrated with the national government-wide M&E structure. Hence, with the support of other constitutional bodies such as the Public Works Department, the CIDB exercises oversight responsibility to ensure the growth, development and sustenance of the South African construction industry. That is to say that the CIDB monitors, evaluates and sanctions non-compliance industry players.

iv Like some construction industries across developed nations, the SA construction industry with the collaboration of the CIDB, PWD and the Centre for Scientific and Industrial Research (CSIR) have instituted construction industry indicators (CIIs) as performance standards specific to the construction industry and these are reviewed consistently. These CIIs are measures for the compliance of all stakeholders, suppliers, contractors, clients and consultants.

Summary

This chapter discussed M&E from the perspective of developing nations with the focus on Kenya and South Africa. The practice of M&E in both economies was shown to be well instituted and implemented in the public sectors of the economy. This chapter also presented an overview of the construction industry in these countries. M&E regulatory policies guiding the implementation of M&E and the major setbacks facing the implementation of M&E in the countries were discussed. Finally, findings and lessons learnt were also espoused. The next chapter will focus on the review of the Ghanaian construction industry and the practice of M&E.

References

Abrahams, M. A. (2015). A review of the growth of monitoring and evaluation in South Africa: Monitoring and evaluation as a profession, an industry and a governance tool. *African Evaluation Journal*, 3(1), pp 1–8.

Auma, E. (2014). Factors affecting the performance of construction projects in Kenya: A survey of low-rise buildings in Nairobi Central Business District. *The International Journal of Business & Management*, 2(12), p. 115.

Basheka, B. C. & Byamugisha, A. (2015). The state of monitoring and evaluation (M&E) as a discipline in Africa. *African Journal of Public Affairs*, 8(3), pp. 75–95.

Basson, R. (2013). South Africa: South African Monitoring and Evaluation Association (SAMEA). Voluntarism, consolidation, collaboration and growth. The case of SAMEA. In: Rugh, J. and Segone, M. (Eds.). *Voluntary Organizations for Professional Evaluation (VOPEs): Learning from Africa, Americas, Asia, Australasia, Europe and Middle East*. UNICEF, pp. 262–274.

Bowen, P., Edwards, P. & Cattell, K. (2012). Corruption in the South African construction industry: A mixed methods study. In: Smith, S. D. (Ed.). *Proceedings of the 28th Annual Association of Researchers in Construction in Management (ARCOM) Conference*, Edinburgh, UK: ARCOM, pp. 521–531.

Construction Industry Development Board (CIDB) (2014). *Annual Report, 2013/2014*.

Construction Industry Development Board (CIDB) (2016b). *Annual Report, 2015/2016*.

Construction Industry Development Board (CIDB) (2017a). *Construction monitor, Transformation*. Q4, 2016.

Construction Industry Development Board (CIDB) (2017b). *Compliance monitor, Register of projects*. Q4, 2016.

Dassah, M. & Uken, E. (2006). Monitoring and evaluation in Africa with reference to Ghana and South Africa. *Journal of Public Administration*, 41(4), pp. 705–720.

Department of Provincial and Local Government (2006). *Municipal infrastructure. Roles and responsibilities of national sector departments, provincial counterparts and municipalities*. Department of Provincial and Local Government.

Engela, R. & Ajam, T. (2010). *Implementing a government-wide monitoring and evaluation system in South Africa.* Washington, DC: Independent Evaluation Group, World Bank.

Gitahi, K. K. (2015). *Determinants influencing monitoring and evaluation processes of road construction projects in Kenya National Highways Authority (KeNHA), Central Region, Kenya.* Master's dissertation. Kenya: University of Nairobi.

Kamau, C. G. & Mohamed, H. B. (2015). Efficacy of monitoring and evaluation function in achieving project success in Kenya: A conceptual framework. *Science Journal of Business and Management,* 3(3), p. 82, doi:10.11648/j.sjbm.20150303.14

Kenya. Ministry of National Planning and Development and Vision 2030 (MPND). (2012). *National Monitoring and Evaluation Policy.*

Khumalo, J., Mashiane, J. & Roberts, S. (2014). Harm and overcharge in the South African precast concrete product cartel. *Journal of Competition Law and Economics,* 10(3), pp. 621–646, doi:10.1093/joclec/nhu005

Kiarie, A. W. & Wanyoike, D. (2016). Determinants of successful implementation of government-funded projects in Kenya: A case study of integrated financial management information system. *International Journal of Innovative Research and Development,* 5(10), pp. 169–175.

Marx, J. H. (2014). *Results of the 2014 survey of the CIDB construction industry indicators.* Construction Industry Development Board.

McGuckin, S. (2017). *Global construction industry: a new normal? International Construction Market Survey: Building momentum.* Turner & Townsend.

Mertens, D. M. & Russon, C. (2000). A proposal for the International Organization for Cooperation in Evaluation. *American Journal of Evaluation,* 21(2), pp. 275–283.

Ministry of Devolution and Planning (2016). *Guidelines for the development of county integrated monitoring and evaluation system.* Government of the Republic of Kenya.

Mtshali, Z. (2015). *A review of the monitoring and evaluation systems to monitor the implementation of early childhood development within Gauteng department of health.* Master's dissertation, Stellenbosch University, Stellenbosch.

Mugambi, F. & Kanda, E. (2013). Determinants of effective monitoring and evaluation of strategy implementation of community-based projects. *International Journal of Innovative Research and Development,* 2(11).

Mugo, P. M. & Oleche, M. O. (2015). Monitoring and evaluation of development projects and economic growth in Kenya. *International Journal of Novel Research in Humanity and Social Sciences,* 2(6), pp. 52–63.

Mulama, K., Liguyani, P. & Musiega, D. (2014). Effectiveness of monitoring and evaluation in enhancing performance of social development projects in Busia County – A survey of government social development projects. *International Journal of Management Research and Reviews,* 4(8), pp. 773–796.

Mwangi, J. K., Nyang'wara, B. M. & Ole Kulet, J. L. (2015). Factors affecting the effectiveness of monitoring and evaluation of Constituency Development Fund projects in Kenya: A case of Laikipia West Constituency. *Journal of Economics and Finance,* 6(1), pp. 74–87.

Mwangu, A. W. & Iravo, M. A. (2015). How monitoring and evaluation affects the outcome of Constituency Development Fund projects in Kenya: A case study of projects in Gatanga Constituency. *International Journal of Academic Research in Business and Social Sciences,* 5(3), doi:10.6007/IJARBSS/v5-i3/1491, pp 13–31

Nene, N., (2015). 2015 Budget Speech, South African National Treasury, viewed 10 March 2017, from www.treasury.gov.za/documents/national%20budget/2015/speech/speech.pdf.

Njama, A. W. (2015). *Determinants of effectiveness of a monitoring and evaluation system for projects: A case of Amref Kenya WASH programme.* University of Nairobi.

Odero, W., Reeves, W. and Chokerah, J. (2017). African Economic Outlook: Kenya 2017. AfDB, OECD and UNDP.

Ofori, G. (2000). Challenges of construction industries in developing countries: Lessons from various countries. *In: 2nd International Conference on Construction in Developing Countries: Challenges Facing the Construction Industry in Developing Countries,* Gaborone, November, pp. 15–17.

Ogolla, F. & Moronge, M. (2016). Determinants of effective monitoring and evaluation of government-funded water projects in Kenya: A case of Nairobi County. *Strategic Journal of Business & Change Management,* 3(1). pp 329–358

Onjure, C. O. & Wanyoike, D. M. (2016). Influence of monitoring and evaluation practices on performance of national-government-funded construction projects in Uasin Gishu County – Kenya. *International Journal of Innovative Research and Development,* 5(12), pp. 78–95.

Pillay, P. & Mafini, C. (2017). Supply chain bottlenecks in the South African construction industry: Qualitative insights. *Journal of Transport and Supply Chain Management,* 11(0), doi:10.4102/jtscm.v11i0.307. pp 1–12

Report Buyer (2016). *Construction in South Africa – Key trends and opportunities to 2020.*

SA Construction (2013). *Highlighting trends in the South African construction industry.* South Africa: PricewaterhouseCoopers (PwC).

SA Construction (2014). *Highlighting trends in the South African construction industry* 2nd edition. South Africa: PricewaterhouseCoopers (PwC).

SA Construction (2015). *Highlighting trends in the South African construction industry* 3rd edition. South Africa: PricewaterhouseCoopers (PwC).

SA Construction (2016). *Highlighting trends in the South African construction industry* 4th edition. South Africa: PouseCoopers.

Statistics South Africa, (2016). Manufacturing and construction industries report 2016, viewed 05 April 2017, from http://www.statssa.gov.za/?page_id=624

Tengan, C. & Aigbavboa, C. (2016). Evaluating barriers to effective implementation of project monitoring and evaluation in the Ghanaian construction industry. *Procedia Engineering,* 164, pp. 389–394, doi: 10.1016/j.proeng.2016.11.635.

The Central Intelligence Agency (2016). The World Fact Book 2016. Retrieved from https://www.cia.gov/library/publications/download/download-2016/index.html

Wachaiyu, V. W. (2016). *Monitoring and evaluation factors influencing success of development projects: A case of Starehe Sub-Country, Kenya.* Master's dissertation, University of Nairobi, Kenya.

Wanjira, J. (2016). Kenya's construction industry and its challenges. Construction Review Online, Available March 20, 2017 from: https://constructionrevie-wonline.com/2016/09/kenyas-construction-industry-and-its-challenges/.

Windapo, A. O. & Cattell, K. (2013). The South African construction industry: perceptions of key challenges facing its performance, development and growth. *Journal of Construction in Developing Countries,* 18(2), p. 65.

Windapo, A. O. (2016). Skilled labour supply in the South African construction industry: The nexus between certification, quality of work output and shortages. *SA Journal of Human Resource Management,* 14(1), pp. 1–8.

World Economic Situation and Prospects (2014). *Country classification: Data sources, country classifications and aggregation methodology.*

11 A review of the Ghanaian construction industry and the practice of monitoring and evaluation

11.1 Abstract

The chapter was structured to review the monitoring and evaluation practices of the Ghanaian construction industry. The review informs that the National Development Planning Commission (NDPC) is mandated to lead and coordinate the preparation of development plans and commence the monitoring and evaluation of the country's development effort. This mandate is subsequently decentralized to the district planning coordinating unit at the MMDAs with a composition of nearly 11 team members. Further, to ensure M&E guarantees project success, routine activities such as project supervision from start to finish, i.e. site visits and inspection, certifying project quality, progress reporting and periodic site meetings should be undertaken. The study further avers that stakeholder involvement in the construction project M&E is limited to the three key stakeholders identified as the client, contractor and the consultant. It is therefore imperative to consciously and constantly manage knowledge of all kinds during the M&E of the project.

11.2 Introduction

Ghana is a West African country boarded in the east by Togo, in the south by the Gulf of Guinea (Atlantic Ocean), in the north by Burkina Faso and in the west by Cote D'Ivoire and it covers an area of 238,500 km^2 or 92,085 m^2. Map 11.1 shows the map of Ghana, indicating her surrounding neighboring countries with international borders, the ten-regional capitals with regional boundaries, major cities, roads and the airport, with Accra being the national capital. According to the Ghana Statistical Service (2012), the population of the country as per the 2010 population and housing census stood at 24,658,823 with an annual intercensal growth rate of 2.5%. However, currently (August 2017), the CIA Fact Book (2016) revealed an annual population growth rate of 2.18%. In 2010, the Greater Accra region according to the Ghana Statistical Service (2012) was the most densely populated (1,236/km^2) region owing to the rural-urban migration for better living conditions.

The country is endowed with natural resources such as gold, cocoa, diamond, bauxite and, recently, oil and gas which serves as the main foreign exchange

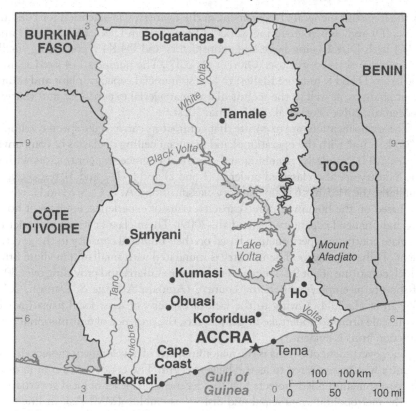

Map 11.1 Map of Ghana.

earner and, sometimes, budgetary support from grants and loans from the international community for major infrastructure development. Ghana is currently a
low middle-income country based on the World Bank's categorization.

11.3 Overview of the Ghanaian construction industry

The Ghanaian construction industry (GCI), unique as it may be, shares similar
characteristics with other construction industries in developing countries. These
characteristics may include bureaucracy, financial uncertainty, unregulated labor
market and poor project management practices (Ahadzie, 2007) resulting from
weak technological, structural and economic conditions prevailing in other developing countries (Ofori, 2000). The industry is dominated by the activities of main
or specialist contractors, private or public clients who are mostly the financiers of
the projects and consultants who ensure projects are delivered to the right specification, cost and within the set time. The Ministry of Water Resources Works and
Housing (MWRWH) and the Ministry of Transport (MoT) are responsible for
contractor classification after companies have been incorporated to do business

by the Registrar General's Department in the country. Classification for building works (D) and civil engineering works (K) ranges from D1K1, D2K2, D3K3 and D4K4 with D1/K1 being large-scale contractors and D4/K4 representing small- and medium-scale contractors (Ofori et al., 2017). The numbers 1–4 fixed against the letters D and K indicate highest to lowest financial capacity, plant and equipment holdings, as well as the technical and managerial expertise of staff (Laryea & Mensah, 2010; Ofori et al., 2017; Sena, 2012).

These classifications are to ensure that contractors can execute specific volumes of work in line with the operational and financial ceiling of classes of companies (Sena, 2012). Similarly, plumbing and electrical engineering contractors within the industry are also classified under G (1 and 2) and E (1, 2 and 3), respectively. Similarly, the MoT has 6 main category classifications: A, B, C, S, M and L which are based on the human resource capacity, years of experience, equipment holding and financial capability (Agbodjah, 2009). These classifications are strict to regulate contractors' performance based on their financial capacity in the government of Ghana projects. The industry is inundated with small and medium firms which constitute about 90% of registered firms in Ghana and providing over 80% of short-term employment in the country (Amoah, Ahadzie & Dansoh, 2011; Tengan et al., 2014). Owing to the lack of capacity of most local firms (mostly small-scale firms) to undertake bigger projects, the presence of multinational construction firms is evident.

The government of Ghana is the major financier of development projects in the country which is referred to as the biggest client. The concentration of most of the government-funded projects is at the decentralized level of local governance, i.e. the metropolitan, municipal and district assemblies (MMDAs). In line with government's biggest interest, the Ghana Highway Authority (GHA), the defunct State Construction Corporation (SCC) and the Architectural Engineering Service Limited (AESL) were established by the government to manage the formal construction sector (Agbodjah, 2009). Subsequently, other state organizations have established departments and units in their setup for similar purposes such as the planning and works department of MMDAs as well as private organizations. To ensure fairness in the award of government/public projects, a regulatory framework, namely the Public Procurement Act 663 of 2003 was promulgated. The Act provides for public procurement, establishes the procurement boards and makes administrative and institutional arrangements for procurement and stipulates the tendering procedure (Public Procurement Act 663, 2003).

11.3.1 Significance of the Ghanaian construction industry

The construction industry (CI) continues to play a key role in the socio-economic conditions and the built environment in Ghana (Osei, 2013). The industry provides social, educational, health and economic infrastructure to drive economic growth. It designs, builds, repairs and demolishes all kinds of building, civil, mechanical and electrical engineering works in the economy (Ofori, 1980). The CI's contribution to the GDP and other indicators of the economy has been

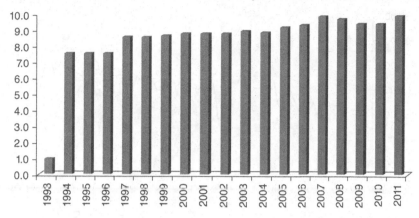

Figure 11.1 Relative share of the construction sector to GDP.
Source: Osei, 2013

extensively reported (Agbodjah, 2009; Laryea & Mensah, 2010; Ofori, 1980). According to Anvuur, Kumaraswamy and Male (2006), the annual volume of goods and services procured for the year 2003 stood at an approximate 10% of the GDP. Figure 11.1 also illustrates the GCI's contribution to the GDP and the overall industry output by 9.1% and 35.9%, respectively between the period 1993 and 2011 (Osei, 2013).

The construction industry is also acknowledged for affecting many other sectors of the economy. The construction industry depends on materials from the manufacturing sector such as the cement producers and roofing companies (Osei, 2013). Darko and Löwe (2016) citing Sutton and Kpentey (2012) revealed the height of cement usage in the construction industry. It is indicated that cement use increased from 4.8 million metric tons in 2010 to 5.5 million metric tons in 2012, representing an increase in consumption of about 15% in nearly two years (Darko & Löwe, 2016). Additionally, Osei (2013) reports the heavy reliance of the construction sector on the automobile industry for the provision of transport and site equipment such as cranes, the painting and coating sector as well as the textile and the clothing industry for the provision of special protective wear for the industry's consumption. The volume of short- to long-term employment created cannot be overstated, particularly where the industry is largely manual. Ofori (2012) revealed that the construction industry in all countries contributes between 5% and 10% to the GDP and employs about 10% of the working population. A trend outlook of cement production in Ghana is illustrated in Figure 11.2 which shows a steady rise in cement production, an indication of growth in the manufacturing sector due to the demand of the construction sector.

Ofori (2012) informs that the inadequate knowledge of the strategic role of the construction industry has perhaps affected the industry in contributing fully to the Ghanaian economy. Government is the major player in the construction business and so to achieve the full potential of the construction industry,

Figure 11.2 Trend outlook of cement production in Ghana.

Source: Osei (2013)

government particularly must give priority to the industry; unfortunately, not the case according to Anaman and Osei-Amponsah (2007) in promoting economic growth and development.

11.3.2 Challenges of the Ghanaian construction industry

Just as other sectors of the economy, the construction industry is inundated with several categories of challenges. Three categories of challenges to the Ghanaian construction are discussed, namely industry challenges, economic challenges as well as project management and implementation challenges.

11.3.2.1 Industry challenges

The Ghanaian construction industry is described as fragmented and like many others in developing nations, this makes administration of the industry difficult. This is due to the fact that several aspects of the industry are under the purview of different ministries, making coordination difficult, hence the need for a unified body to administer, coordinate and regulate the practices of the industry (Ofori, 2012). The industry's lack of capacity (financial, human resource, technology) to undertake complex projects (Ofori, 2012) has caused an influx of multinational firms from across the globe who have the technology and expertise to undertake complex construction projects (Laryea, 2010). Further, Ofori (2012) laments at the poor level of technology and knowledge transfer (TKT) in the Ghanaian construction industry. He acknowledges that technologies and knowledge on construction methods are growing at a much faster pace, hence the need for industry practitioners to learn and adapt to emerging technologies for greater efficiency of the industry.

There is a lack of partnership between the industry and government which represents the major partner of the development of the country (Anaman & Osei-Amponsah, 2007). Capital supply constraints of the industry appear to be a major challenge yet to be tackled. Several scholars, including Fugar, Ashiboe-Mensah and Adinyira (2013), Ofori (2012) and Ofori-Kuragu, Owusu-Manu and Ayarkwa (2016) have described the non-existence of an industry development board or council as an impediment to the growth of the industry in Ghana. This board or council with an exclusive mandate to regulate the construction industry practice along with other recognized professional bodies would be a step towards industry growth. Indeed, Ofori (1980) avers the high prevalence of poor performance of the construction industry in developing countries is primarily as a result of a lack of a regulatory organization. This lack of a regulatory body in the industry was further stressed by Ephson (2017) in a newspaper article in which he provides an overview of the Ghanaian construction industry with perspectives on the challenges the industry faces – unfortunately, this has resulted in unprofessional practices by unlicensed professionals in the industry (Ahmed, Hatira & Valva, 2014).

11.3.2.2 Economic challenges

"High unstable business environment" is the description given to the Ghanaian construction industry by many scholars, particularly Ofori-Kuragu et al. (2016, p. 132). The construction industry in Ghana is characterized by high inflation, interest rate volatility, price fluctuation and in recent times, the poor power supply **(dumso)** has been studied regarding its dire consequences on the performance of the construction industry. Amoah, Ahadzie and Dansoh (2011) reported on the speculation by industry players about the inability of the government to create an enabling fiscal environment for the industry to thrive, knowing that the majority of firms are small- and medium-scale. For instance, Chileshe and Yirenkyi-Fianko (2012) evaluated risk factors affecting the Ghanaian construction industry. The study involved twenty-five risk factors as ranked by consultants, contractors and clients in the GCI. Inflation and price fluctuation were found as two economic risk factors which were ranked second and fourth, respectively out of the twenty-five risk factors of the industry (Chileshe & Yirenkyi-Fianko, 2012).

In the real estate sector of the industry, Amos, Gadzekpo and Amankwah (2015) revealed challenges in housing finance which are attributed to causes such as unstable inflation, the depreciation of the cedi and high interest rates. The Bank of Ghana (2007) further laments the unstable macroeconomic environment and weak legal and regulatory environment as limiting the mortgage market in Ghana.

11.3.2.3 Project management and implementation challenges

Ghana's development agenda is hinged on the local government system towards promoting infrastructure development in the country (Osei-Kyei & Chan, 2016). To facilitate fair and even distribution of development across the country, the decentralization system of governance was born at the instance of the promulgation of the Local Government Act, 1993 (Act 462) and provided for by the 1992 Constitution of Ghana [Local Government Act, 1993 (Act 462)]. The local government levels, also referred to as the Metropolitan, Municipal and District Assembly (MMDA), have since been the fulcrum of development, receiving technical and financial support from the central government (CG), civil society organizations (CSOs) and donor agencies (DA) towards development. The main functions outlined in the Local Government Act, 1993 (Act 462), section 10, sub-sections 3a, i and ii entrust the responsibility of the overall local government development to the MMDA and through the Regional Coordinating Council (RCC), prepare and submit development plans of the MMDA and accompanying budgets to the National Development Planning Commission (NDPC) and the Minister for Finance respectively for approval, among other responsibilities, all geared towards infrastructure and economic development of the MMDA [Local Government Act, 1993 (Act 462)].

The infrastructure development process of the MMDAs is realized through the construction of health, educational and economic infrastructure project delivery. Construction project delivery (CPD) at the MMDA level has enormous benefits to the constituents or their citizens. The delivery of construction projects engages the majority of the small-scale building contractors (SSBCs) in the country who are noted to create jobs for the local community (Amoah, Ahadzie & Dansoh, 2011). Hence, to ensure projects initiated achieve the outlined objectives, project performance measurement is critical and this will be realized through effective monitoring and evaluation. In the management and implementation of projects, numerous challenges are encountered which are predominantly contractor- and process-related. Laryea (2010) reviewed extant literature regarding the challenges and opportunities facing contractors in Ghana and revealed that contractors face financial challenges due to delayed payment with no compensation. They also lacked managerial and technical capacity, there is a high level of competition resulting in an inability to secure jobs and the contractors have a poor attitude towards repayment of secured loans. Anvuur et al. (2006) also revealed the poor procurement practices arising from long post-award negotiations, prolonged evaluation processes and a dispute over land for development in the Ghanaian construction industry which has led to the insecurity of construction project funding. Additionally, in furthering the debate regarding the lack of transparency and corruption in the procurement process, Ameyaw, Mensah & Osei-Tutu, (2012) revealed dire procurement practices as posing a great challenge to the industry. These factors he outlines as low capacity levels of professionals on procurement, poor interaction between procurement bodies and the public procurement authority, non-adherence to procurement laws and the deliberate manipulation of the procurement process rendering the process non-competitive (Ameyaw et al., 2012). Osei-Tutu et al. (2010) explored the corrupt related challenges in the construction industry which need immediate attention if the procurement process is to achieve its relevance in the industry. They found bid rigging, price differentials, collation, fraud and bribery as major corrupt procurement challenges faced in the construction industry. Similarly, Asamoah and Decardi-Nelson (2014) identified practices such as embezzlement, tender manipulation, favoritism and corruption which they describe as unethical practices in the Ghanaian construction industry.

Poor workmanship has been an age-old challenge in the Ghanaian construction industry. Baiden and Tuuli (2004) describe as persistent the problem of construction defects in the industry. Danso (2014) underscores the reality of poor workmanship in the Ghanaian construction industry. He further mentions the factors promoting poor workmanship and reports on the poor supervision and the use of the adulterated material as a major precursor to the poor workmanship challenge faced in the GCI. The challenge of poor workmanship is attributed to the poor competencies of labor and the wrong tools and equipment used on site which describes the level of competency, skill shortage and lack of human resources in the Ghanaian construction industry.

As reported earlier in this study, the Ghanaian construction industry is made up of nearly 90% of small- and medium-scale buildings and civil engineering contractors. This category of firms is confronted with managerial capacity challenges as reported by Amoah et al. (2011) when they studied the factors affecting construction performance in Ghana, highlighting the views of small- and medium-scale buildings and civil engineering contractors. Similarly, Ofori-Kuragu, Baiden and Badu (2016) report on the perceived inadequate knowledge of contractors in the application of the necessary management techniques in terms of the poor performance of the Ghanaian construction industry. Djokoto, Dadzie and Ohemeng-Ababio, E. (2014) and Ametepey, Aigbavboa and Ansah, (2015) in separate studies on barriers to sustainable construction in Ghana revealed a poor attitude towards change regarding construction methods and material use, hence the slow pace of the industry's adoption of sustainable construction practices. Similarly, Fugar et al. (2013) studying the implication of the capital theory postulated the inability to embrace change in the GCI as a serious challenge. A recent study by Ametepey, Gyadu-Asiedu and Assah-Kissiedu (2017) on the cause-effect relationship of delayed local government projects in Ghana suggested that schedule delay is a common occurrence in the Ghanaian construction industry posing considerable challenges to the industry. These delays inevitably caused litigation and increased the final project cost. Additionally, Fugar et al. (2013) identified an insufficient financial resource, low level of worker mobility, the non-existence of appreciation of the workforce, the absence of investment in human resource development and low technology as dire challenges confronting the construction industry in Ghana.

11.4 Construction project monitoring and evaluation practice in Ghana

The overarching challenge with the construction industry in Ghana has been that of poor performance while the impact of M&E in uplifting performance has been widely reported. Project management processes, drawing on the M&E of inputs and processes have been described as a major approach to ensure project management is effective and that projects are successfully implemented (Otieno, 2000). Idoro (2012) further describes M&E as providing checks and balances for ensuring that the strategies and overall project objectives are achieved.

In Ghana's construction industry, the practice of M&E can be described from two main approaches, namely the formal and the informal settings of project implementation. The informal project implementation allows private project developers to engage professionals to develop and manage projects till successful completion. Professionals design and execute the project, ensuring that projects are implemented to meet the standard quality specification, within clients' budget and that projects are delivered on schedule through constant supervision and monitoring of progress to ensure the efficient use of project resources (materials, funds and labor). Professionals also find strategies to mitigate challenges influencing negative construction practices. In the formal setting of project implementation, government ministries, department and agencies such as the Ghana Education Trust Fund (GETFund), the MWRWH, the metropolitan, municipal and district assemblies (MMDAs), the Ministry of Education and Health roll out infrastructural projects to solve the infrastructure need of constituents and the country at large. The services to ensure the projects are executed to set targets are done in-house. However, in most cases, these services are subcontracted to project management consultants (PMCs) owing to project locations, lack of capacity and logistics to effectively manage these projects.

11.4.1 Project management consultants

Project management consultants (PMCs) are a team of professionals (quantity surveyor, architect, all relevant categories of engineers, planners) with the requisite knowledge, skills, tools and techniques to manage project activities to ensure project requirement is met. Sarda and Dewalkar (2016) posit that the role of PMCs in civil engineering project includes scheduling, risk identification, cost budgeting, monitoring and control by project plans to ensure project success. In Ghana, Ahadzie, Proverbs and Olomolaiye (2004) indicate the steady interest in involving project managers since the late 1980s on projects with the aim of achieving success. The need for and relevance of effective M&E integrating project management consultants in the Ghanaian construction industry cannot be overemphasized and, as such, is critical for project success.

In small-scale projects, for instance, project implementation units (PIUs)/ technical department (TD) of the MMDA, ministries, departments and agencies (MDAs) may undertake project management services. However, in the case of large-scale projects (project scope), the functions of PM consultants become necessary for the efficient and effective management of projects. Other factors such as project requirements, financial threshold limitations provided in the procurement of works and services and limited financial and technical capacity compel the procurement of project consultants.

11.4.2 Ministry of water resources works and housing (MWRWH)

It is worth noting that to ensure successful project implementation, work activities and building and civil engineering contractors must be taken seriously, hence their regulation in the construction industry (Ofori-Kuragu et al., 2016). The task of classifying contractors is the business of the MWRWH (see section 5.3 for details of contractor classifications in Ghana). The formulation and implementing policies, strategies and programmes for the Housing and Works sub-sectors of Ghana are vested in the MWRWH (2010). The Ministry serving as a central management agency (CMA) and overseeing several departments with the specific task and mandate in ensuring project implementation is successful under two broad agencies, namely the works sector and the housing sector agencies. The works sector's agency has the Hydrological Services Department (HSD) which is concerned with the programming and the coordination of coastal protection works, as well as the construction and maintenance of storm drains countrywide and the monitoring and evaluation of surface water bodies in respect of floods.

Also, the Public Works Department (PWD) established in 1850 by the colonial government is the oldest government department and is responsible for overseeing the construction and maintenance of all government bungalows, office blocks and other landed properties. Further, the Architectural and Engineering Services Limited (AESL) found in all ten administrative regions of the country was established to provide consultancy services primarily to government departments, (para-statal) organizations as well as individuals. They also can undertake surveys and the design of bridges, irrigation works, sewage systems, water supplies, soil and foundation studies as well as the testing of construction materials and the valuation and appraisal of properties. The service provided by AESL also includes the supervision of civil and building works and, finally, the provision of quality control services to contractors. The Engineers' Council is also represented under the Ministry. The housing sector agencies of the MWRWH also include the Rent Control Department (RCD), Department of Rural Housing (DRH), Public Servants Housing and Loans Scheme Board (PSHLSB), the Architects' Registration Council (ARC), Tema Development Corporation (TDC) and the State Housing Company (SHC).

11.4.3 Ghana Education Trust Fund (GETFund)

The Ghana Education Trust Fund (GETFund) was established by an Act of Parliament, Act 581 of 2000. The Act defines the core mandate of GETFund to include the provision of funding to complement efforts to improve educational infrastructure in public educational institutions. Additionally, funding to procure educational equipment and staff development is provided for in the Act. Regrettably, just as other project implementation agencies, GETFund projects were saddled with several forms of challenges such as delays in honoring payment certificates and poor and ineffective monitoring. Studies have shown incessant delays by the Ministry of Finance and Economic Planning (MoFEP) in transferring GETFund's allocations as scheduled. This explains the delays in honoring payment certificates to contractors engaged in GETFund projects (Banaman, 2015). This unfortunate delay is a barrier to effective M&E.

A study by Eyiah-Botwe (2015) cited a report titled "GETFund Review and Outlook (2000–2009)" of the GETFund consultative forum held in 2010 which attributed the poor management of resources to the absence of M&E on the project. Hence, the technical department (TD) was subsequently tasked with the responsibility of identifying, appraising and supervising all projects initiated by the GETFund. Specifically, the TD is tasked with construction and engineering professionals responsible for managing GETFund projects across the country, coordinating functions, the M&E of project activities to assess project outcomes and effectiveness and identifying improvements for future projects. The Technical Department also handles the procurement of works and services under the funding, providing contract administration and infrastructure needs assessment (INA) for present and future funding options.

11.4.4 *Metropolitan, municipal and district assemblies (MMDAs)*

To ensure uniform development across the entire country, the local government structure was promoted. Specifically, the developmental function at the local government level is vested in the metropolitan, municipal and district assemblies (MMDAs) and is provided for by the Local Government Act, 1993 (Act 462) under section 10, sub-section 3 (a)-(e) to include the following:

a Responsibility for the total development of the district and guaranteeing the preparation and submission through the regional coordinating council;

   i development plans of the district to the National Development Planning Commission for approval; and

   ii the budget of the district related to the approved plans to the Minister responsible for Finance for approval;

b Formulating and implementing plans, programmes and policies for the effective deployment of the needed resources for the general development of the district;

c Encouraging and supporting productive activity and social development in the district and removing any obstacles to initiative and development;

d Initiating plans for the development of the basic infrastructure and providing municipal works and services in the district; and

e Developing, improving and managing human settlements and the environment in the district.

Given the development functions of the local government, the District Planning Coordination Unit was established under the Planning Department to enforce all planning functions related to all development activities in line with the NDPC framework [Local Government Act, 1993 (Act 462)]. This function is successfully achieved with the partnership of the Works and Engineering Department of the MMDAs.

11.5 Monitoring and evaluation regulatory policy in Ghana

The NDPC is the policy regulatory body created under Articles 86 and 87 of the 1992 Constitution of the Republic of Ghana. Following the constitutional provision, the National Development Planning Commission Act, 1994 (Act 479) and the National Development Planning (System) Act, 1994 (Act 480) were enacted by the Parliament of Ghana in establishing the NDPC. Articles 86 and 87 of the 1992 Constitution of the Republic of Ghana clearly delineate the composition and functions of the NDPC which direct them to lead and coordinate the preparation of development plans and to commence the M&E of the country's development efforts (NDPC, 2008). The results-based M&E system (RBMES) and results-based budgeting (RBB) is the adopted approach by the NDPC to monitor, evaluate and co-ordinate development policies, programmes and projects to ensure cost-effectiveness, institutional capacity strengthening, promotion of good governance, accountability and credibility to the partners and government as provided under the National Development Planning Act, 1994 (Act 479), section 2(2)(e) (NDPC, 1994a). The RBME approach, however, focused on the outcome and impact of the implementation of the programme or project.

The mandate of the NDPC was decentralized to the newly created decentralized development planning systems in Ghana, thus the district planning coordination unit (DPCU) at the MMDAs (NDPC, 1994b). As part of the coordination of the planning and development services of the local governance structure, the NDPC provides an external M&R service to the MMDAs. Key stakeholders are required to undertake effective M&E of projects (Tengan & Aigbavboa, 2017a), hence an 11-member team is constituted [Local Government Act, 1993 (Act 462)] and may co-opt any other sector agencies and persons from both the private sector and civil society organizations with relevant expertise in a given area (NDPC, 2008). The DPCU was subsequently established under section 46 of the Local Government Act, 1993 (Act 462). Under the Ghana Poverty Reduction Strategy (GPRS II), the DPCU in fulfilling the M&E function undertakes the following:

i Being directly responsible for the development and implementation of the District M&E Plan;
ii Convening quarterly DMTDP performance review meetings with all stakeholders. It is essential that representatives of the NDPC and RPCU attend the quarterly meetings;
iii Undertaking periodic project site inspections;
iv Liaising with the RPCU to agree on goals and targets;
v Defining indicators for measuring change, especially on gender equity and other cross-cutting themes in GPRS II, such as vulnerability, exclusion and social protection;

vi   Collecting and collating feedback from the sub-district levels for preparation of the District APR;

vii Facilitating dissemination and public awareness creation on GPRS II, the Annual Progress Reports and other documents from NDPC at district and sub-district levels;

viii Providing support to GSS to undertake district-level CWIQ and other national surveys and census;

ix   Producing District Annual Progress Reports and making recommendations for policy review; and

x   Conducting mid-term and terminal evaluations of the DMTDP.

From the foregoing, there is evidence to suggest that government ministries, departments and agencies (MDAs) such as the GETFund and the Architectural and Engineering Services Limited (AESL) have structured technical departments and units mandated to undertake the M&E of all construction and development projects initiated. The AESL reports on their M&E plan covering resources and activities through the MWRWH. However, inadequate vehicular logistics and delays in the release of funds were outlined as the major challenges of the agency in delivering its M&E mandate (MWRWH, 2010). Also, the AESL and the PWD of the MWRWH outline the functions of the department towards project delivery. Hence, ensuring projects meet quality, cost and time are critical performance criteria for these technical departments to achieve.

The Local Government Service (2014) provides the professional responsibilities of individual professional staff outlined in the scheme of service in the case of the MMDAs to strengthen and delineate duties towards achieving coordinated efforts in the project delivery. For instance, the DPCU of the MMDA is responsible for leading in strategic planning, efficient integration and implementation of public policies and programmes to achieve sustainable economic growth and development as well as efficiently controlling the resources and ensuring that field activities are efficiently performed to produce the desired output. The engineering unit also formulates policies for the efficient management and administration of the project and provides technical backstopping for the regional coordinating councils and MMDAs, in addition to offering technical assistance to the Local Government Service Secretariat, Regional Coordinating Councils and the MMDAs in all engineering matters (Local Government Service, 2014). Regarding the architectural unit the scheme of service of the local government requires them to advise management and provide inputs for the formulation and implementation of architectural policies; undertake site inspections to guide and advise on architectural projects; prepare designs and approve architectural drawings on projects; and prepare modules of structures (Local Government Service, 2014).

A documentary review of the M&E practices at MDAs and MMDAs in Ghana suggests the existence of regulatory policies and frameworks on M&E. The NDPC provides plans for the monitoring and reporting of projects and programmes of all the relevant sectors, departments and agencies that received funding from the central government for their project and programme implementation. What is not clear is the existence of an established policy framework and guideline at these ministries, departments and agencies and the metropolitan, municipal and district assemblies on how M&E of construction projects should be done to achieve quality, schedule and cost targets as well as the project impact on the beneficiary community. At best, the existing plan considers the monitoring of development projects to conform to permit acquisition and building regulation adherence and expenditures of resources (budget) towards programme implementation.

11.6 Implementation of monitoring and evaluation systems in the Ghanaian construction industry

This section considers the M&E system implementation in the Ghanaian construction industry which is aimed at achieving project objectives such as ensuring quality of the projects and their delivery within the allocated time and budget. Other targets of M&E implementation on projects include ensuring that the project activities and the project outcome are satisfactory and impact positively on the beneficiary community. In the Ghanaian construction industry, to guarantee that M&E of the projects results in achieving project quality, cost and time, activities such as constantly undertaking project supervision from inception to completed and ready to be handed over; thus, site visits and inspection, certifying project quality, progress reporting and periodic site meetings are important. Three key M&E activities, namely site visit and inpection (supervision), meetings and reporting are discussed in the ensuing sections.

11.6.1 Site visit and inspection (supervision)

Site visit and inspection are vital in project implementation undertaken either at the request of the contractor to resolve some challenges (Wong, 2005) or as a scheduled activity for the project. M&E requires stakeholders, particularly the project consultants, project implementation units or agencies such as the MMDAs and the clients (donors/financiers) to visit project sites to observe the progress and conformity to agreed conditions, the contract and the works. It is however imperative that during such site visits and inspections, any instruction given at the site that seeks to influence the scope, cost and quality of the project is to be issued through the designated consultant rather than the project consultant or his representative. The client and project implementation unit or agency report on their observations to the consultant for redress with the contractor. This is to ensure that all key stakeholders' interest is incorporated to ensure satisfaction. The consultant, on considering the challenges and the volume of the task involved in the project, may station a clerk of works on the project to ensure their representation on the project is constant to facilitate effective project M&E. The inspection on the project affords the M&E team to reveal the bottlenecks and challenges in the project. Methods of construction, material specification and workmanship are brought under strict scrutiny. During the inspection, certain quality standards are confirmed through laboratory tests such as the concrete cube test while other tests are conveniently conducted at the site. The consultant also issues instructions during site visits and inspections to enforce compliance with specifications and change orders (variations) to some part of the work which may or may not have cost implications.

11.6.2 Site meeting

Site meetings are organized with the main aim of ensuring the smooth implementation of a construction project to meet specific project targets. Gorse and Emmitt (2009) acknowledge the importance of progress meetings. During M&E, an opportunity is provided to bring together all the parties (stakeholders) to the project to a meeting where project implementation challenges and successes are discussed. This ensures the relationship with project stakeholders is maintained and sustained for effective project delivery (Gorse & Emmitt, 2003). Oke, Mavimbele and Aigbavboa (2016) postulate that site meetings stand out as the oldest as well as current means of dealing with problems stemming from construction projects and enforcing the acceptable standard (project specifications). These scheduled meetings may be an agreed periodic meeting with some intermittent emergency meetings. Before such site meetings which are usually organized at the project site office, stakeholders visit and inspect the project. The contractor presents his revised works programme and progress report highlighting lags in the programme and measures to mitigate such a programme. The cash flow position on the project, the physical progress against expected progress and similar issues are reviewed to ascertain whether the project is progressing within budget and cost while quality reports from laboratory tests are presented for the benefit of all stakeholders.

Several decisions are made in respect of how to address all the challenges on the project and to ensure delivery of the project to the project specifications. Stakeholders are empowered; they become more aware of the state of the project and have a feeling of ownership of the project. Hence, the capacities of stakeholders on M&E are enhanced. Further, Oke et al. (2016) maintain that site meetings help in enforcing project quality standards such as developing a project quality control plan, assessing workmanship during construction, increasing communication in the construction team and assessing the specification used. Potential disputes and claims that may arise are prevented (Love et al., 2010).

11.6.3 Progress reporting

Projects fail for several reasons. As opined by KPMG (2014), a key consideration for project success, yet one that is difficult to undertake, is effective project reporting. Project reporting involves communicating information regarding the duration of project time utilized, the quality level of the project and the cash flow status of the project against the planned targets. Resources available at the site (material and labour) are reported as well as the physical volume of work done. Reporting M&E findings (progress) to key stakeholders is therefore critical. The M&E team, most likely led by the project consultant/manager, reports project progress and findings to all key stakeholders to the project, particularly project donors and clients. This is to ensure observations made during the site visit and inspection and agreed decisions at site meetings and site instructions regarding the modification to the project are communicated officially to all parties to the project.

Construction projects generate enormous amounts of information (Fong & Chu, 2006) and therefore it is necessary to process, store and disseminate such information to all parties involved in the project. However, systems employed in reporting progress have been sporadic in the construction industry. According to Craig and Sommerville (2006), storage, disseminating and managing of project information stop with each of the individual project team members involved in the project. Thus, the quantity surveyors keep to themselves the cost data information generated by them on the project while the architects and the engineers alone hold the architectural and structural design information. Participatory M&E facilitates progress reporting which is anchored in the learning and improvement of project delivery, meeting reporting requirements of donor agencies and also in facilitating knowledge management on project delivery.

11.7 Stakeholders involved in M&E in the Ghanaian construction industry

The Ghanaian construction industry thrives on the involvement and participation of stakeholders in project delivery. Thus, clients, contractors, consultants and donors come together in a project environment. Without stakeholders, there would be no industry. Scholarly works have categorized stakeholders into three main segments; those who affect the project, those who will be affected by the project and, finally, those who may be interested in the outcome of the project (Mathur, Price, Austin & Moobela, 2007). For instance, project stakeholders have been defined as those who can potentially influence the implementation of a project (Freeman, 2004; Fewings, 2005). Also, Tengan and Aigbavboa (2017) assert that stakeholders are those individuals and groups who benefit directly or indirectly from the outcome of a project. Further, Bourne (2010) and the project management institute (PMI, 2013) refer to stakeholders as individuals, groups of persons or organizations who are perceived to affect or are affected by decisions, activities or outcomes of a project.

The influence of stakeholders, therefore, is imperative in the M&E of construction projects. Their participation and involvement in M&E are critical to safeguarding their interest (Tengan & Aigbavboa, 2017). Guijt, Arevalo and Saladores (1998) maintain that participatory M&E is cost-effective, provides accuracy and relevant information and is empowering while the World Bank posits that the active participation of primary stakeholders in M&E provides the opportunity for building the capacity of stakeholders. In a study to understand the level of engagement and participation of stakeholders in the monitoring and evaluation of projects in the Ghanaian construction industry, Tengan and Aigbavboa (2017) identified seven stakeholders in the delivery of projects. However, their involvement in the M&E of projects was limited to the three key stakeholders who affect the project. These stakeholders are identified as the client, contractor and the consultant (Tengan & Aigbavboa, 2017). The study further revealed that material suppliers and local authority service providers were not critical participants in M&E but may affect the project while the beneficiary community largely provided the labor for the delivery of the project, indicating their benefit in the project (Tengan & Aigbavboa, 2017).

11.8 Barriers to effective construction project monitoring and evaluation implementation in Ghana

The need to execute projects to achieve success (performance) has been admitted as significant, just as the implementation process. When the implementation process is not effective and efficient, projects might be delivered but not to any performance requirement, hence the failure of the project. In the Ghanaian construction industry, M&E has been entrenched in institutional structures to monitor and evaluate developmental projects. The challenge therefore with the overarching task of poor performance is discussed in this section.

Most construction projects in the formal environment are initiated through funding from central government (GoG) and other government departments and ministries. For instance, government funds educational infrastructure (laboratories and classrooms), health infrastructure (CHIPS compounds and hospitals), social infrastructure (markets) and industrial infrastructure (factories) through the GETFund, the Ministries of Health and Education and budgetary allocations to the MMDAs. There also exist other sources of donor funding for initiating projects at the local government level which include the District Development Fund (DDF) and the Urban Grant (UG). Literature has indicated that there is a substantial delay in the process of the allocation and the release of funds for smooth projects implementation. Effective M&E is therefore hampered when projects are not smoothly implemented owing to lack of timely and adequate funding. Tengan and Aigbavboa (2016) assert that limited financial resources and budgetary allocation affect the implementation of M&E at the MMDA level. Capacities of M&E implementation departments or units have also been recognized as a barrier to the effective implementation M&E of construction projects. Similarly, while project stakeholders are engaged in the project delivery, their participation in the M&E of public projects at the local government level was very poor (Tengan & Aigbavboa, 2017).

Planning for M&E is critical for its effectiveness and efficiency during implementation. M&E tasks are geared towards specific performance measures and, as such, the inputs, processes and indicators need to be planned well. What, how and who to monitor and evaluate are essential to measure the effectiveness and efficiency of M&E and the performance of the monitored project. The lack of planning due to the absence of M&E systems on GETFund projects resulted in incessant delays on projects (Eyiah-Botwe, 2015). Institutional structures and capacity for M&E or to support M&E at the local government level influence the effective and efficient M&E implementation. According to Tengan and Aigbavboa (2016), metropolitan, municipal and district assemblies in Ghana have a weak institutional capacity which affects the effective M&E of projects. This probably explains the many incomplete and abandoned projects initiated by MMDAs across the country (Williams, 2015).

11.9 Knowledge management in the monitoring and evaluation of projects

The construction industry, according to Carillo and Anumba (2002), has been described as a knowledge-intensive sector. Likewise, knowledge has been defined to mean the use of information by persons involved in a project, utilizing their skills, technical competencies, understandings, opinions, commitment and motivations (Mugula, 2015). The theory on knowledge management (KM) suggests there is a common ground for understanding what KM entails. Whereas Davenport and Prusak (1998) aver that KM involves the process of capturing, distributing and effectively utilizing knowledge, Robinson et al. (2005) indicate that KM is the process of creating, acquiring, capturing, sharing and using knowledge. Similarly, KM has been described as the process of developing, preserving, using and sharing knowledge to enhance organizational learning and performance (Mugula, 2015).

KM has been studied extensively in many economic sectors for its benefit and influence in achieving organizational performance and to establish a competitive advantage over other organizations (Omotayo, 2015). In the construction industry, however, knowledge management practice has been sporadic and low owing to the fragmented nature, scattered knowledge over different processes and a wide number of trades involved in the industry (Hashim, Talib & Alamen, 2014). In the M&E of projects in the construction industry, stakeholders with different interest and knowledge are drawn together to deliver a project to achieve organizational and project objectives such as enhancing project performance, creating a competitive advantage, innovations, sharing of lessons learned, integration and continuous improvement (Mugula, 2015). Individual stakeholders' knowledge on M&E implementation is therefore brought to bear on the projects which, in most cases, are not aligned to each other and can possibly cause delays due to differences in understanding by stakeholders on how M&E should be done. It is therefore imperative to consciously and constantly manage knowledge of all kinds during the M&E of the project.

Summary

This chapter presented an overview of the Ghanaian construction industry and the practice of M&E of construction project delivery. Specifically, an appreciation of the geographical positioning, structure and characteristics of Ghana were presented. Also discussed were the characteristics of the Ghanaian construction industry, its significance and relevance in national development and the challenges of the industry. It was evident from the review that the Ghanaian construction sector appreciates the importance of M&E in project implementation. Ministries, departments and agencies (MDAs) linked with project implementation have established units in charge of M&E. At the local government level, the Planning Unit and the Works Department complement each other to ensure the effective implementation of projects initiated. However, challenges such as lack of technical capacity of M&E staff, low utility of M&E findings, inadequate budget allocation and limitation of project stakeholders in M&E implementation were clear since projects consistently failed. The review of key stakeholders involved in the M&E of construction project revealed contractors, consultants and clients as key actors in the M&E while others such as material suppliers and local authority service providers were not critical participants in M&E but may affect the project while the beneficiary community largely provided the labor for the delivery of the project, indicating their benefit in the project. Effective M&E knowledge management is therefore imperative. The next chapter using the Ghanaian construction industry as a case study adopted the Delphi methodology to establish the need for and relevance of M&E in the construction industry, the main and sub-attributes that bring about effective M&E and whether the attributes are comparable to other country settings, identify the critical challenging factors that influence M&E and finally to determine the impact of effective M&E on construction project delivery.

References

Agbodjah, L. S. (2009). *A human resource management policy development (HRMPD) framework for large construction companies operating in Ghana.* Kumasi, Ghana: Kwame Nkrumah University of Science and Technology.

Ahadzie, D. K. (2007). *A model for predicting the performance of project managers in mass house building projects in Ghana.* PhD Thesis, UK: University of Wolverhampton.

Ahadzie, D. K., Proverbs, D. G. & Olomolaiye, P. (2004). Meeting housing delivery targets in developing countries: The project manager's contribution in Ghana. *Globalisation and Construction*, pp. 619–630.

Ahmed, K., Hatira, L. & Valva, P. (2014). *How can the construction industry in Ghana become sustainable?* Karlskrona, Sweden: Blekinge Institute of Technology.

Ametepey, O., Aigbavboa, C. & Ansah, K. (2015). Barriers to successful implementation of sustainable construction in the Ghanaian construction industry. *Procedia Manufacturing*, 3, pp. 1682–1689, doi:0.1016/j.promfg.2015.07.988

Ametepey, S. O., Gyadu-Asiedu, W. & Assah-Kissiedu, M. (2017). Causes-effects relationship of construction project delays in Ghana: Focusing on local government projects. In: Charytonowicz, J. (Ed.). *Advances in intelligent systems and computing.* Presented at the International Conference on Human Factors, Sustainable Urban Planning and Infrastructure, California, USA: Springer, pp. 84–95.

Ameyaw, C., Mensah, S. & Osei-Tutu, E. (2012). Public procurement in Ghana: The implementation challenges to the Public Procurement Law, 2003 (Act 663). *International Journal of Construction Supply Chain Management*, 2(2), pp. 55–65.

Amoah, P., Ahadzie, D. K. & Dansoh, A. (2011). The factors affecting construction performance in Ghana: The perspective of small-scale building contractors. *The Ghana Surveyor*, 4(1), pp. 41–48.

Amos, D., Gadzekpo, A. & Amankwah, O. (2015). Challenges of real estate development in Ghana from the developers' perspective. *Developing Country Studies*, 5(10), pp. 66–75.

Anaman, K. A. & Osei-Amponsah, C. (2007). Analysis of the causality links between the growth of the construction industry and the growth of the macro-economy in Ghana. *Construction Management and Economics*, 25(9), pp. 951–961, doi:10.1080/01446190701411208

Anvuur, A., Kumaraswamy, M. & Male, S. (2006). Taking forward public procurement reforms in Ghana. In: *Proceedings of the 2006 CIB W107: Construction in Developing Countries International Symposium: Construction in Developing Economies: New Issues and Challenges*. CIB.

Asamoah, R. O. & Decardi-Nelson, I. (2014). Promoting trust and confidence in the construction industry in Ghana through the development and enforcement of ethics. *Information and Knowledge Management*, 3(4), pp. 63–68.

Baiden, B. K. & Tuuli, B. (2004). Impact of quality control practices in sandcrete blocks production. *Journal of Architectural Engineering*, 10(2), pp. 53–60.

Bank of Ghana (2007). *The housing market in Ghana*.

Bourne, L. (2010). Why is stakeholder management so difficult? In: *Congresso International*, Columbia: Universidad EAN Bogota.

Carillo, P. M. & Anumba, C. J. (2002). Knowledge management in the AEC sector: An exploration of the mergers and acquisition context. *Knowledge and Process Management*, 9(3), pp. 149–61.

Chileshe, N. & Yirenkyi-Fianko, A.B. (2012). An evaluation of risk factors impacting construction projects in Ghana. *Journal of Engineering, Design and Technology*, 10(3), pp. 306–329, doi:10.1108/17260531211274693

Craig, N. & Sommerville, J. (2006). Information management systems on construction projects: Case reviews. *Records Management Journal*, 16(3), pp. 131–148, doi: 10.1108/09565690610713192.

Danso, H. (2014). Poor workmanship and lack of plant/equipment problems in the construction industry in Kumasi, Ghana. *International Journal of Management Research*, 2(3), pp. 60–70.

Darko, E. & Löwe, A. (2016). *Ghana's construction sector and youth employment*. London: Overseas Development Institute. Working Paper.

Davenport, T. & Prusak, L. (1998). *Working knowledge: How organisations manage what they know*. Boston, MA.: Harvard Business School Press.

Djokoto, S. D., Dadzie, J. & Ohemeng-Ababio, E. (2014). Barriers to sustainable construction in the Ghanaian construction industry: Consultants' perspectives. *Journal of Sustainable Development*, 7(1), doi:10.5539/jsd.v7n1p134.

Ephson, B. (2017). Overview of the construction industry in Ghana. *The Daily Dispatch*, 2 May 2017.

Eyiah-Botwe, E. (2015). An evaluation of stakeholder management role in GETFund polytechnics projects delivery in Ghana. Available online at: www.iiste.org, 7(3), pp. 66–73.

Fewings, P. (2005). *Construction project management: An integrated approach*. USA and Canada: Taylor & Francis.

Fong, P. S. & Chu, L. (2006). Exploratory study of knowledge sharing in contracting companies: A sociotechnical perspective. *Journal of Construction Engineering and Management*, 132(9), pp. 928–939.

Freeman, R. E. (2004). The stakeholder approach revisited. *Zeitschrift für Wirtschafts-und Unternehmensethik*, 5(3), pp. 228–241.

Fugar, F. D. K., Ashiboe-Mensah, N. A. & Adinyira, E. (2013). Human capital theory: Implications for the Ghanaian construction industry development. *Journal of Construction Project Management and Innovation*, 3(1), pp. 464–481.

Ghana Statistical Service (2012). *2010 Population and housing census, final results*. Accra.

Gorse, C. A. & Emmitt, S. (2003). Investigating interpersonal communication during construction progress meetings: Challenges and opportunities. *Engineering, Construction and Architectural Management*, 10(4), pp. 234–244, doi:10.1108/09699980310489942

Gorse, C. A. & Emmitt, S. (2009). Informal interaction in construction progress meetings. *Construction Management and Economics*, 27(10), pp. 983–993, doi:10.1080/01446190903179710

Guijt, I., Arevalo, M. & Saladores, K. (1998). Participatory monitoring and evaluation. *PLA Notes*, 31, p. 28.

Hashim, E. M. A. B., Talib, N. A. & Alamen, K. M. (2014). Knowledge management practice in Malaysian construction companies. *Middle-East Journal of Scientific Research*, 21(11), pp. 1952–1957, doi:10.5829/idosi.mejsr.2014.21.11.21727.

Idoro, G. I. (2012). Influence of the monitoring and control strategies of indigenous and expatriate Nigerian contractors on project outcome. *Journal of Construction in Developing Countries*, 17(1), p. 2012.

Laryea, S. (2010). Challenges and opportunities facing contractors in Ghana. In: Laryea, S., Leiringer, R. and Hughes, W. (Eds.). *Procs West Africa Built Environment Research (WABER) Conference*, 27–28 July, 2010, Accra, Ghana, pp. 215–226.

Laryea, S. & Mensah, S. (2010). The evolution of indigenous contractors in Ghana. In: Laryea, S., Leiringer, R. and Hughes, W. (Eds.). *Presented at the West Africa Built Environment Research (WABER) Conference*. Accra, Ghana: West Africa Built Environment Research, pp. 579–588.

Local Government Service (2014). *Local government scheme of service* (Revised edition).

Love, P., Davis, P., Ellis, J. & Cheung, S. O. (2010). Dispute causation: identification of pathogenic influences in construction. *Engineering, Construction and Architectural Management*, 17(4), pp. 404–423, doi:10.1108/09699981011056592

Mathur, V. N., Price, A. D., Austin, S. A. & Moobela, C. (2007). Defining, identifying and mapping stakeholders in the assessment of urban sustainability. In: Horner, M., Hardcastle, C., Price, A., and Bebbington, J. (Eds). *Presented at the International Conference on Whole Life Urban Sustainability and its Assessment*, Glasgow.

Ministry of Water Resources Works and Housing (MWRWH). (2010). Monitoring and evaluation plan and budget for 2010–2013.

Mugula, R. (2015). M&E enhances knowledge management and organizational learning in public sector. *The New Times*,.

National Development Planning Commission (NDPC) (1994a). *National development planning commission act, 1994, Act 479.*

National Development Planning Commission (NDPC) (1994b). *National development planning (system) act, 1994 Act 480.*

National Development Planning Commission (NDPC) (2008). *Growth and poverty reduction strategies (GPRS II)*. Accra, Ghana.

Ofori, G. (1980). *The construction industries of developing countries: The applicability of existing theories and strategies for their improvement and lessons for the future; the case of Ghana.* England: University of London.

Ofori, G. (2000). Challenges of construction industries in developing countries: Lessons from various countries. Conference Papers. In: *2nd International Conference on Construction in Developing Countries: Challenges Facing the Construction Industry in Developing Countries.* Botswana: Gaborone, November 15–17.

Ofori, G. (2012). Developing the construction industry in Ghana: The case for a central agency. *A concept paper prepared for improving the construction industry in Ghana. National University of Singapore.*

Ofori, P. A., Twumasi-Ampofo, K., Danquah, J. A., Osei-Tutu, E. & Osei-Tutu, S. (2017). Investigating challenges in financing contractors for public sector projects in Ghana. *Journal of Building Construction and Planning Research,* 5(2), pp. 58–70, doi:10.4236/jbcpr.2017.52005

Ofori-Kuragu, J. K, Baiden, B. & Badu, E. (2016). Performance measurement tools for Ghanaian contractors. *International Journal of Construction Management,* 16(1), pp. 13–26, doi:10.1080/15623599.2015.1115245.

Ofori-Kuragu, J. K., Owusu-Manu, D. G. & Ayarkwa, J. (2016). The case for a construction industry council in Ghana. *Journal of Construction in Developing Countries,* 21(2), pp. 131–149, doi:10.21315/jcdc2016.21.2.7

Oke, A., Mavimbele, B. & Aigbavboa, C. (2016). Site meeting as a sustainable construction tool. *Socioeconomica,* 5(9), pp. 83–92, doi:10.12803/SJSECO.590009

Omotayo, F. O. (2015). Knowledge management as an important tool in organisational management: A review of literature. *Library Philosophy and Practice,* p. 1.

Osei, V. (2013). The construction industry and its linkages to the Ghanaian economy – policies to improve the sector's performance. *International Journal of Development and Economic Sustainability,* 1(1), pp. 56–72.

Osei-Kyei, R. & Chan, A. P. (2016). Implementing public–private partnership (PPP) policy for public construction projects in Ghana: Critical success factors and policy implications. *International Journal of Construction Management,* pp. 1–11, doi:10.1080/15623599.2016.1207865

Osei-Tutu, E., Badu, E. & Owusu-Manu, D. (2010). Exploring corruption practices in public procurement of infrastructural projects in Ghana. *International Journal of Managing Projects in Business,* 3(2), pp. 236–256, doi:10.1108/17538371011036563

Otieno, F. A. O. (2000). The roles of monitoring and evaluation in projects. In: *2nd International Conference on Construction in Developing Countries: Challenges Facing the Construction Industry in Developing Countries,* pp.15–17.

Project Management Institute (PMI). (2013). *Managing change in organizations: A practice guide.* UK: PMI.

Public Procurement Act, 2003 (Act 663) of the Republic of Ghana.

Robinson, H. S., Carrillo, P. M., Anumba, C. J. & Al-Ghassani, A. M. (2005). Knowledge management practices in large construction organisations. *Engineering, Construction and Architectural Management,* 12(5), pp. 431–445, doi:10.1108/ 09699980 510627135.

Sarda, A. & Dewalkar, S. (2016). Role of project management consultancy in construction. *International Journal of Technical Research and Applications,* 4(2), pp. 317–320.

Sena, A. A. (2012). Are contractor classifications in Ghana accurate? *The Quantity Surveyor,* 5(1), pp. 4–10.

Sutton, J. and Kpentey, B. (2012). *An enterprise map of Ghana,* London: International Growth Centre.

Tengan, C. & Aigbavboa, C. (2016). Evaluating barriers to effective implementation of project monitoring and evaluation in the Ghanaian construction industry. *Procedia Engineering*, 164, pp. 389–394, doi: 10.1016/j.proeng.2016.11.635.

Tengan, C. & Aigbavboa, C. (2017). Level of stakeholder engagement and participation in monitoring and evaluation of construction projects in Ghana. *Procedia Engineering*, 196, pp. 630–637, doi:10.1016/j.proeng.2017.08.051

Tengan, C., Anzagira, L. F., Kissi, E., Balaara, S. & Anzagira, C. A. (2014). Factors affecting quality performance of construction firms in Ghana: Evidence from small-scale contractors. *Civil and Environmental Research*, 6(5), pp. 18–23.

The Central Intelligence Agency (2016). The World Fact Book 2016. Retrieved from https://www.cia.gov/library/publications/download/download-2016/index.html

Williams, M. (2015). *Project delivery and unfinished infrastructure in Ghana's local governments*. International Growth Center. Policy brief No. 89105.

Wong, K.Y. (2005). Critical success factors for implementing knowledge management in small and medium enterprises. *Journal of Industrial Management and Data Systems*, 105(3), pp. 261–279.

Part V

Insight from Delphi research study

A case of Ghanaian experts

12 Case study

12.1 Abstract

A two-stage Delphi methodology was adopted to establish the need for and relevance of monitoring and evaluation (M&E) in the construction industry, the main and sub-attributes that bring about effective M&E and whether the attributes are comparable to other country settings, to identify the critical challenging factors that influence M&E and, finally, to determine the impact of effective M&E on construction project delivery. Ninety-eight attributes categorized into seven main factors achieved consensus at the end of the second round of the Delphi study. It is also confirmed from the study that M&E practice in the Ghanaian construction industry is influenced by the involvement of key stakeholders, the M&E budget allocation, the approach to M&E implementation, communication and the leadership role. Consensus was achieved on some challenges outlined amongst the 11 experts empanelled. Regarding the relationship between M&E and project success, experts largely agreed that there is a significant relationship suggesting effective M&E will lead to project success.

12.2 Introduction

A case study approach employing the Delphi methodology was undertaken to understand the relevance, determinants, challenges and impact of M&E in construction project delivery. The following Delphi-specific objectives (DSO) were established for the case study chapter of this book:

DSO1   To ascertain the need for and relevance of M&E in the Ghanaian construction industry;

DSO2   To determine the main factors and sub-attributes that bring about effective M&E and to examine whether the attributes that determine effective M&E in other countries are similar to construction project M&E in Ghana;

DSO3   To identify the critical challenging factors that influence M&E in the Ghanaian construction industry; and

DSO4   To determine the impact of effective M&E-determining factors on the success of construction project delivery in Ghana.

The philosophy regarding the above objectives was to arrive at a common consensus on the factors and attributes of effective M&E in the Ghanaian construction industry. The DSO is intended to help settle on the key factors and attributes critically significant to determine effective M&E in Ghana and to help develop a holistically integrated monitoring and evaluation model for the Ghanaian construction industry for implementation by other industries that share similar characteristics.

An expert panel of 20 construction industry and academic professionals with M&E expertise consented and were constituted initially for the first round of the Delphi study. However, 13 experts constituting 65% responded and participated in the first round of the Delphi study. Subsequently, 12 experts responded to the second round of the Delphi study. The major selection criteria for experts compulsorily included experts having consulted or been involved in construction project delivery at the metropolitan, municipal and district assembly (MMDA) level in Ghana, their years of experience in the construction industry and their academic qualification. The criteria were considered significant to the study since experts were required to possess a thorough understanding of the M&E practices at the local government level in the Ghanaian construction industry. Expert panel members were also considered based on their affiliation to recognized professional bodies in the Ghanaian construction industry, namely the Ghana Institution of Surveyors (GhIS), Ghana Institute of Architects (GIA) and the Ghana Institution of Engineers (GhIE). Experts were also empanelled from the academic and research environment as well as those in industry practice in Ghana.

Round One of the questionnaire was designed based on the summaries from a comprehensive review of literature highlighting the main and sub-attributes that are potential in influencing effective M&E implementation in the Ghanaian construction industry. Subsequently, the Delphi Round Two (final round) questionnaire was designed based on the previous response from Round One. Round One of the Delphi study was significant to the study as it served as a brainstorming exercise to empirically include the list of main and sub-attributes that determined effective M&E in the Ghanaian construction industry.

To accomplish this objective therefore closed and open-ended questions were used in Round One after which they were analyzed and formed the basis for Round Two of the study. Frequencies were obtained to measure the level of consensus reached amongst experts regarding the main and sub-attributes that determined effective M&E in the Ghanaian construction industry. The essence of Round Two of the Delphi study was to afford experts the opportunity to revise their earlier responses in Round One of the Delphi study and also respond to the new attributes which were proposed by other experts in Round One and which have the potential to influence effective M&E in the Ghanaian construction industry. After analyzing the response from the second round of the Delphi study, the characteristics and features which determined effective M&E as agreed to by the expert panel were reorganized to present a holistic picture of the attributes that determine effective M&E in the Ghanaian construction industry.

A successful Round Two presented results indicating a general agreement (consensus) amongst experts; a third round of the Delphi study was, therefore, not necessary. The median, mean and standard deviation and interquartile deviation (IQD) scores of each attribute were calculated. In situations where the score was 1 point from the median score, experts were requested to explain their responses. Upon reaching a consensus after the second round of the Delphi study, experts were communicated with regarding the consensus reached. Consensus is said to be achieved with 100% of experts in agreement. However, two-thirds of experts agreeing are considered a common consent (Stitt-Gohdes & Crews, 2004). Good consensus was therefore reached with each question or statement analyzed.

12.3 Demographic characteristics of experts

A list of project managers, academics and industry professionals who were believed to have significant knowledge and expertise on M&E of construction projects was identified from the Ghanaian construction industry. An initial number of 20 experts consented to participate; hence, 20 Delphi questionnaires were sent out for the Delphi Round One. Thirteen responded to Round One while 11 returned Round Two. An attrition rate of 45% was recorded. This represents an 85% response rate. However, a cumulative 55% response rate was achieved over the initial 20 experts identified for the study.

The educational levels of experts indicate 73% had a Master's degree while 27% had a PhD degree in construction-related fields of study. Experts for the study belonged to various recognized professional bodies in the country. Specifically, about 46% were professional members of the Ghana Institution of Surveyors, while representatives of the GIA and the GhIE constituted 27% each.

Experts had substantial years of experience of working in the construction industry. About 46% of experts had between 6 and 10 years' working experience in the construction industry. A total of 27% of the experts also had between 11 and 15 years' experience while 27% had over 16 years of experience in the industry. All experts had been involved in construction project delivery at the metropolitan, municipal and district assembly (MMDA) level in Ghana, either as consultants or builders (contractors). Finally, experts were drawn from diverse institutions in the construction industry. Lecturers at universities constituted about 46%, with research institutions and the construction industry represented by 27% each.

12.4 Delphi-specific objectives

The literature reviewed revealed sets of attribute and sub-attributes which may influence effective M&E implementation in the Ghanaian construction industry. The identified attributes used to determine effective M&E implementation in the Ghanaian construction industry cut across both developed and developing countries and from other fields such as agriculture and health. A factor or attribute rated high/low influence or agreement based on the extent to which the listed attribute may affect M&E implementation in the Ghanaian construction industry. An ordinal scale rating from 1 to 10 (1 = "low influence" or "strongly disagree" or "low impact" and 10 = "high influence", "strongly agree" or "high impact") was utilized. The levels of influence and impact were therefore obtained as the product of the achievement of consensus. Experts were provided with the group median response and their response from the Delphi Round One study. This was to afford experts the opportunity to review their ratings if they were convinced to do so as follows, namely to accept the group MEDIAN, maintain their ORIGINAL response or indicate a NEW response. Also, experts were encouraged to provide reasons when their responses were 10% or one unit above or below the group median response on the agreement, influence and impact scale and also to rate new issues identified from the Delphi Round One study.

Consensus here has been categorized into three scales: strong consensus, good consensus and weak consensus. The scales are presented as follows:

i   Strong consensus: median 9–10, mean 8–10, IQD ≤ 1 and ≥ 80% (8–10)
ii   Good consensus: median 7–8.99, mean 6–7.99, IQD ≥ 1.1 ≤2 and ≥ 60% ≤ 79% (6–7.99)
iii   Weak consensus: median ≤ 6.99, mean ≤ 5.99 and IQD ≥ 2.1 ≤ 3 and ≤ 59% (5.99)

The impact scale was also categorized as follows:

i   VHI: 9 – 10
ii   HI: 7 – 8.99
iii   MI: 3 – 6.99

12.4.1 The need for and relevance of M&E in the Ghanaian construction industry

The need for and relevance of M&E in the Ghanaian construction industry are determined. The following five questions were asked in this regard:

i   Is M&E a relevant project management tool in achieving project success in the Ghanaian construction industry?
ii   Does the Ghanaian construction industry have an M&E Policy Framework document for construction project M&E?
iii   Does the Ghanaian construction industry require an M&E Policy Framework to guide M&E practice in construction project delivery?
iv   Should an M&E Policy Framework be made a responsive criterion for the selection of project consultants and contractors in the GCI?
v   Should organizations establish a separate M&E unit to monitor and report on project implementation?

From Table 12.1, out of the five questions asked, four issues were agreed upon with three of them achieving consensus by experts recording median ratings between

Table 12.1 Availability, need for and relevance of M&E

Questions	Median	Mean	Std. Dev.	IQD
M&E is a relevant project management tool in achieving project success in the Ghanaian construction industry.	8.0	8.09	0.30	0.0
The Ghanaian construction industry DOES NOT have an M&E Policy Framework document for construction project M&E.	10.0	9.36	2.11	0.0
The Ghanaian construction industry requires an M&E Policy Framework to guide M&E practice in construction project delivery.	9.0	8.73	0.9	0.0
An M&E Policy Framework document should be made a responsive criterion for selecting project consultants and contractors in the GCI.	9.0	9.00	3.61	1.5
Organizations should establish a separate M&E unit to monitor and report on project implementation.	9.0	8.45	1.13	1.0

9 and 10 and the IQD between 0 and 1 (IQD ≤ 0.0), respectively. Though it was agreed amongst experts for an M&E policy framework document to be made a responsive criterion for the selection of project consultants in the GCI recording a median score of 9.0 which indicates high agreement among experts, consensus was not achieved based on the IQD score of 1.5. On the other hand, experts averagely agreed that M&E was a relevant project management tool in achieving project success in the Ghanaian construction industry based on the median score of 8.0, but reached a unanimous consensus when it recorded an IQD score of 0.0 and a standard deviation score of 0.30.

Findings suggest that M&E is significantly needed and relevant for implementation in the GCI. This position was found to be consistent with the need for M&E in other industries and the cultural context for achieving project success (Barasa, 2014; Kamau & Mohamed, 2015). Specifically, on the five main questions asked to answer the Delphi study question 1, experts agreed that M&E is a relevant project management tool in achieving project success in the Ghanaian construction industry. This position was found to support the study by Kamau and Mohamed (2015) when they studied the efficacy of M&E functions in achieving project success in Kenya. Further, experts unanimously agreed that the Ghanaian construction industry lacks an M&E policy framework to guide the M&E implementation of projects and, therefore, its importance in supporting the achievement of project success.

The first step to institutionalize M&E is to establish an M&E unit at organizational and institutional levels in Ghana where infrastructure development is targeted and supported with funding from central government and donor agencies. The establishment of M&E units at the institution or organizational level to monitor and report on project implementation was expressly agreed by experts for the study. The need to establish M&E units at organizations is supported by the study by Abrahams (2015) to facilitate close supervision to ensure project success. In a different observation, even though the use of an M&E policy framework as a responsive criterion in selecting project consultants (M&E specialists) and contractors in the GCI was highly agreed upon by experts, obtaining a median score of 9.0, consensus was not achieved for such requirements to be executed as it recorded an IQD of 1.5 which is above the cut-off for the study. Hence, the variableness amongst experts on the statement is explained in the high standard deviation score of 3.61.

12.4.2 The main factors and sub-attributes that determine effective M&E in the Ghanaian construction industry and its relationship with M&E determinants in other countries

A total of 14 factors were used to measure the determinants of M&E in the Ghanaian construction industry. Out of the 14 factors, only 5 factors were ranked as high influence by experts based on the median score range of 9.0–10.0. These factors are budgetary allocation and logistics, technical capacity and training, effective leadership, effective communication and managerial skills. Accordingly, consensus was reached on the factors as having a significant influence on the effectiveness of M&E (Table 12.2). The high consensus rate is based on the IQD score of between 0 and 1 (IQD ≤ 1).

Additionally, whereas eight other factors were perceived to averagely influence effective monitoring and evaluation in the GCI based on their median rating of between 7.0 and 8.99, consensus was agreed on all the factors by experts. This consensus was based on the IQD score of less than 1 (IQD ≤ 1). The influence of "data quality" on M&E in the construction project delivery had a median score

Table 12.2 Effective M&E factors

Factors	Median	Mean	Std. Dev.	IQD
Stakeholder involvement	8.0	8.00	0.45	0.0
Budgetary allocation and logistics	9.0	8.73	0.90	0.0
Politics	7.0	7.36	1.12	1.0
Technical capacity and training	9.0	8.64	0.81	0.0
Approach to monitoring and evaluation	8.0	7.55	0.82	0.50
Leadership	9.0	9.00	0.45	0.0
Communication	9.0	8.91	0.3	0.0
Organizational culture	7.0	7.00	0.00	0.0
Monitoring and evaluation information systems	8.0	7.73	0.67	0.0
Advocacy	8.0	7.36	1.21	0.50
*Data quality	6.0	6.73	1.01	1.0
*Management skill	9.0	8.09	1.38	1.0
*Relationship between goals and outcome	8.0	7.82	0.75	1.0
*Beneficiary community participation	8.0	7.45	0.93	1.0

* *Indicates new factors introduced from Delphi Round One*

Table 12.3 Stakeholders' involvement attributes

Attributes	Median	Mean	Std. Dev.	IQD
Engaging and participation of stakeholders in M&E	8.0	7.91	0.54	0.0
Providing stakeholder need for M&E	8.0	7.73	0.65	0.0
Taking prompt action on M&E reports and findings	9.0	8.73	0.79	0.5
Recognition of patriotic stakeholders	7.0	7.09	0.54	0.0
Motivating stakeholders towards M&E	8.0	7.73	0.65	0.0
Experience of stakeholders in M&E	8.0	7.82	0.98	0.0
Stakeholder interest in and expectation of M&E	8.0	8.09	0.54	0.0
Identifying stakeholders	8.0	8.00	0.89	0.50
Managing stakeholders' power structures and influence on the project	9.0	8.82	0.40	0.0
Stakeholders' interest and involvement in M&E	9.0	8.91	0.54	0.0
Collaboration at all levels among stakeholders	8.0	8.00	0.45	0.0
Satisfying stakeholders	8.0	7.73	0.65	0.0
Training and developing stakeholders on M&E	8.0	8.00	0.63	0.0

of 6.0, signifying low influence by experts. The factor, however, achieved consensus based on the IQD score of 1.0.

The impact of attributes on the determinants of effective M&E in the Ghanaian construction industry was measured. Thirteen attributes were used to measure the impact of stakeholder involvement on the effectiveness of M&E. Out of the 13 attributes listed, experts perceived three attributes as having a very high impact on M&E (VHI: 9.0–10). With reference to Table 12.3, all remaining ten attributes received a high impact rating (HI: 7.0–8.99) by all experts, suggesting there were none of the attributes found not to have an impact on stakeholder involvement. Subsequently, based on the IQD score, all 13 attributes achieved good consensus (IQD ≤ 1).

Nine budgetary allocation and logistics characteristics were listed to describe their impact on M&E in the GCI. Only one characteristic, "clear budget line for M&E", out of the nine listed characteristics, received very high impact ratings by the experts (VHI: 9.0–10) as shown in Table 12.4. Consensus was, however, not achieved for that

Table 12.4 Budgetary allocation and logistics attributes

Attributes	Median	Mean	Std. Dev.	IQD
Clear budget line for M&E	9.0	8.18	1.33	1.5
Availability of sufficient funds for M&E activities	8.0	8.00	0.45	0.0
Method of budgeting for M&E	7.0	6.91	0.83	0.0
Allocating resource for M&E against the progress of work	8.0	7.73	1.01	0.0
Form and frequency of M&E audit (internal and external)	8.0	7.82	0.75	0.0
Scope and complexity of M&E	8.0	7.55	1.29	0.50
Established M&E units	8.0	7.73	1.10	0.50
Duration of M&E	7.0	6.82	0.98	0.0
Source of funding for M&E	8.0	7.82	0.98	0.50

Table 12.5 Political influence attributes

Attributes	Median	Mean	Std. Dev.	IQD
Political decisions (e.g. spending and budgetary allocations)	9.0	8.18	1.08	1.50
Government policy (e.g. taxation)	8.0	8.09	0.94	0.0
Economic condition (e.g. inflation)	8.0	7.91	0.54	0.0
Change in government	9.0	8.91	0.7	0.0
International relations	8.0	7.45	1.04	0.50
Existing laws and regulations (e.g. procurement law)	8.0	8.00	0.77	0.0
Management interference in M&E	9.0	8.45	1.04	0.50
Source of project funding	8.0	7.91	0.83	0.0
Composition of M&E team	9.0	8.73	0.65	0.0
Favouritism	8.0	7.55	1.13	0.50

attribute (clear budget line) as it obtained an IQD score of 1.5 which is beyond the cut-off (IQD ≤ 1) for the study. Eight other features of budgetary allocation and logistics received high impact ratings from experts with a median score between 7.0 and 8.0 (HI: 7.0–8.99). The IQD score of these features was below the cut-off for the study, thus less than 1 (IQD ≤ 1). Hence, consensus was achieved for all eight features.

Regarding the assessment of political influence on M&E, ten attributes were listed. Four attributes were rated by experts to impact very highly (VHI: 9.0–10). These attributes were political decisions, change of government, management interference in M&E and the composition of the M&E team (Table 12.5). Six other attributes obtained a high impact rating by experts, having recorded a median score of 8.0. All attributes except one (political decision) achieved consensus, thus obtaining an IQD score below 1 (IQD ≤ 1). "Political decision" had an IQD score of 1.5 indicating low consensus (IQD ≥ 1.1 ≤ 2). None of the attributes was found among experts not to have a high impact on M&E in the construction industry in Ghana.

In the assessment of the impact of technical capacity and training on M&E as presented in Table 12.6, 11 characteristics were listed. Experts rated five of the features as having a high impact with a median score of 9.0 (VHI: 9.0–10) while the other features were rated as high impact (HI: 7.0–8.99). Ten of the eleven features achieved high consensus with their IQD rating ranging between 0 and 1 (IQD ≤ 1). The feature "knowledge on M&E" failed to reach consensus because it obtained an IQD score greater than the cut-off for the study, thus 1.50. Experts commented that

> stakeholders and M&E team members do not require prior knowledge of monitoring and evaluation to be effective. The training component and the participation in the M&E will ensure relevant knowledge on M&E. (Anonymous expert).

Subsequently, when the approach to M&E attributes was evaluated, findings revealed that two attributes, namely "use of right M&E tools and techniques"

Table 12.6 Technical capacity and training attributes

Attributes	Median	Mean	Std. Dev.	IQD
Frequency of training	8.0	7.82	0.40	0.0
Content of training	9.0	8.55	1.21	0.0
Planning process of learning intervention	8.0	7.82	0.60	0.0
Educational and training level of M&E team	9.0	8.73	0.90	0.0
Knowledge on M&E	9.0	8.00	1.67	1.50
Method of training	8.0	8.18	0.40	0.0
Mode of training	8.0	8.00	0.63	0.0
The expectation of employees prior to the training	8.0	8.09	0.70	0.0
Involvement in training	9.0	8.55	0.93	1.0
Support for training	9.0	8.73	0.65	0.0
*Desire for training	8.0	8.18	0.87	0.50

* *Indicates new factors introduced from Delphi Round One*

and "frequency of M&E" were rated to have a very high impact (VHI: 9.0–10) on M&E, obtaining a median score of 9.0 each. All other attributes recorded a median score of 8.0, suggesting a high impact rating on M&E (HI: 7.0–8.99). None of the attributes, however, was found to have no impact. Regarding the evaluation for consensus amongst experts, all the attributes recorded an IQD score of 0.0 which signifies high consensus amongst the experts (Table 12.7).

Fifteen attributes were listed to assess the impact of effective leadership on M&E. Amongst the attributes, four were ranked as very high impact (VHI: 9.0–10), obtaining a median score of 9.0 each. Furthermore, 11 attributes recorded median scores between 7.0 and 8.0, signifying a high impact (HI: 7.0–8.99) rating amongst the experts. Table 12.8 presents the interquartile deviation scores of the attributes. One attribute, effective communication, even though was rated to have a very high impact on leadership, did not achieve consensus. It obtained an IQD score of 2.0 which is far above the cut-off for the study (IQD ≤ 1). The remaining attributes, however, achieved consensus, having IQD less than 1.

Table 12.7 Approach to M&E attributes

Attributes	Median	Mean	Std. Dev.	IQD
Use of appropriate M&E tools and techniques	9.0	8.55	1.37	0.0
Frequency of M&E	9.0	9.09	0.3	0.0
Planning for M&E	8.0	8.27	0.65	0.0
Approach to data collection	8.0	8.00	0.89	0.0
Implementation of M&E systems and plans	8.0	8.00	1.18	0.0
Approach to data analysis	8.0	8.00	0.45	0.0
Dissemination of M&E results and findings	8.0	7.91	0.70	0.0
Composition of M&E team	8.0	8.09	0.30	0.0

Table 12.8 Leadership attributes

Attributes	Median	Mean	Std. Dev.	IQD
Leadership style	8.0	8.27	0.90	0.0
Culture and attitude	7.0	7.00	0.89	0.0
Vision	8.0	7.82	0.75	0.0
Commitment	9.0	8.73	0.65	0.0
Personality	8.0	8.09	0.3	0.0
Traits	8.0	7.82	0.6	0.0
Managerial skills	8.0	8.09	0.7	0.0
Gender	8.0	7.73	0.9	0.0
Competencies	8.0	8.00	0.89	0.0
Organizational environment	9.0	9.09	0.3	0.0
*Knowledge	8.0	8.45	0.82	1.0
*Performance	8.0	8.27	0.65	1.0
*Effective communication	9.0	9.00	0.89	2.0
*Behaviour of leader	8.0	8.00	0.63	0.0
*Situation	9.0	8.36	1.03	1.0

* *Indicates new factors introduced from Delphi Round One*

Effective communication as a critical determinant of M&E listed 14 attributes to assess its impact (Table 12.9). Four attributes which include effective communication skills, reporting systems, poor communication structure and the use of the right media recorded a median score of 9.0 each which signifies a very high impact rating (VHI: 9.0–10). Ten other listed attributes recorded a high impact rating (HI: 7.0–8.99). None of the attributes was found not to have an impact. Regarding agreed consensus amongst experts on the attributes, "the use of the right media" even though achieved high impact score of 9.0, consensus was not

Table 12.9 Effective communication attributes

Attributes	Median	Mean	Std. Dev.	IQD
Channel of communication	8.0	8.27	0.79	0.50
Distortion in communication	8.0	7.82	0.87	0.50
Communicator (Sender)	8.0	7.82	0.75	0.0
Intended audience	8.0	7.73	0.65	0.0
Relevance of the communication (Content)	8.0	8.09	0.54	0.0
Effective communication skills	9.0	8.82	0.40	0.0
Appropriate feedback channel	8.0	8.00	0.45	0.0
Access to information	8.0	8.09	0.3	0.0
Reporting systems	9.0	8.45	1.04	0.5
Receiver of the information	8.0	7.82	0.6	0.0
*Proper communication structure	9.0	8.64	0.81	1.0
*Standardization of communication documents	8.0	8.27	0.79	0.5
*Effective listening skills	8.0	8.36	0.81	1.0
*Use of the right media	9.0	8.82	0.98	1.5

* *Indicates new factors introduced from Delphi Round One*

reached since it obtained an IQD score of 1.5 which is above the cut-off rating for consensus for this study. All other attributes recorded an IQD score less than or equal to 1 (IQD ≤ 1). It was, however, clear that the only attribute which did not achieve consensus amongst experts was because it was perceived to be duplication with "the channel of communication". It received a comment from an expert suggesting that:

> use of right media is the same as the channel of communication. (Anonymous expert)

Findings emanating from the Delphi study suggest that the main and sub-attributes that determine effective M&E in GCI are comparable to other countries. Consensus was also achieved on all 14 main attributes that determine effective M&E. They had been found to be strong determinants of M&E in other contexts (Hardlife & Zhou, 2013; Kamau & Mohamed, 2015; Mugo & Oleche, 2015; Musomba et al., 2013; Ogolla & Moronge, 2016; Seasons, 2003). The influence of data quality in determining effective M&E was, however, found to be very weak, obtaining a median score of 6.0 which contradicts the findings of Mulandi (2013).

In the case of the 13 sub-attributes used in assessing the impact of stakeholders' involvement in M&E, again there was agreement on all attributes in significantly influencing stakeholder involvement. All attributes had an IQD rating of less than 1 which is within the cut-off for achieving consensus as set for the study. A high median rating was also observed, ranging between 7.0 and 9.0. Experts' rating suggests the sub-attributes had a high impact on M&E of construction projects in the GCI. Likewise, attributes of budgetary allocation and logistics were rated to have a high impact on M&E, indicating similar characteristics in other contexts and environments. Regarding the clear budget line for M&E, even though it was rated to have an impact, consensus amongst experts was not achieved. The attribute obtained mean and standard deviation scores of 8.18 and 1.33, respectively. The high standard deviation score suggests the variableness of the attributes amongst experts. The findings on these attributes contradict findings of the study by Hwang and Lim (2013) which suggest allocation of funds for M&E was necessary to ensure success. Similarly, while the influence of politics on M&E recorded a very high rating amongst experts in the context of the GCI, political decisions did not achieve consensus even though the attribute was rated to have a very high impact on M&E, based on the median score of 9.0 which corroborates with other studies (Kamau & Mohamed, 2015; Muiga, 2015; Musomba et al., 2013; Seasons, 2003; Waithera & Wanyoike, 2015).

Other attributes such as international relations and favouritism both recorded a high impact on M&E, having obtained a median score of 8.0 each and an IQD of 0.50, but they recorded high standard deviation scores of 1.04 and 1.13, respectively, signifying the variability amongst experts on the attributes. Assessing the sub-attributes of technical capacity and training characteristics, these were classified by experts as having a high impact on determining effective M&E in the

GCI. Consensus was achieved on all attributes except the knowledge on M&E attribute which recorded an IQD score of 1.50 and a standard deviation score of 1.67, indicating high variability. Elsewhere in Kenya, lack of knowledge on M&E was attributed to negatively influencing the success of M&E of donor-funded food security intervention projects (Kimweli, 2013). Even though consensus was not achieved on knowledge on M&E having a significant impact in the Ghanaian construction industry, experts were convinced that the attribute had a very high impact, recording a median score of 9.0.

Regarding the impact of approach to M&E attributes on effective M&E, experts found all attributes to have a high impact on M&E which suggests similar findings from other cultural contexts (Van Mierlo, 2011). Similarly, consensus was achieved on all attributes. However, the use of appropriate tools and techniques saw a standard deviation of 1.37, implying high variableness amongst experts on the attribute. Whereas effective communication is recognized as having a very high impact on leadership in M&E by experts and similar studies by Luthra and Dahiya (2015), it recorded an IQD score of 2.0 rating by experts, revealing no consensus on attributes. All other attributes of leadership were rated to have a high and very high impact (Bikitsha, Mamafha & Ngomane, 2014; Iqbal et al., 2015; Kolzow, 2014; Popa, 2012) based on the observed median scores between 7.0 and 9.0 and an IQD score less than or equal to 1. Studies from other cultural contexts and fields indicate that effective communication had a significant impact on M&E (Mugambi & Kanda, 2013; Windapo, Odediran & Akintona, 2015). The study listed 14 attributes to measure the impact of effective communication on M&E which indicated a high impact rating by experts in the GCI. However, the "use of right media" for communication did not achieve consensus, recording an IQD of 1.50 with a relatively low standard deviation score of 0.90. The non-consensus could be attributed to the perceived duplication of the attribute. Experts were of the view that the "use of the right media" for communication is explained in the effective *communication channel* which measured the high impact on communication on M&E to achieve project success.

12.4.3 Critical challenging factors that influence M&E in the Ghanaian construction industry

The impact of factors posing as challenges to the effective implementation of M&E was also evaluated. Twelve challenges impacting effective M&E in the Ghanaian construction industry were listed to measure their impact by experts in the Delphi study. The assessment revealed five attributes recording a median score of 9.0 each. This high median score suggests a very high impact (VHI: 9.0–10) while the remaining attributes garnered median scores between 7.0 and 8.99, attributing a high impact rating of the challenging attributes to M&E (Table 12.10). In referring to the achievement of consensus, all attributes received 100% consensus, obtaining IQD scores of less than 1 (IQD ≤ 1).

Challenges that have been listed to have a dire impact on M&E in other sectors and cultural contexts such as lack of comparable definition (Patton, 2003),

Table 12.10 Challenges to effective M&E attributes

Attributes	Median	Mean	Std. Dev.	IQD
Time constraints	8.0	7.91	0.94	0.0
Technical capacity and skill of staff	8.0	7.91	0.54	0.0
Financial resource constraints/lack of adequate budget allocation	9.0	8.73	0.47	0.50
Lack of institutional capacity	8.0	7.82	0.40	0.0
Communication challenges	8.0	7.91	0.70	0.0
Lack of sufficient project information	7.0	6.82	0.40	0.0
Inconsistencies in project design, specifications, etc.	8.0	8.00	0.77	0.0
Lack of M&E unit	9.0	8.45	0.69	1.0
Lack of M&E plan	8.0	7.91	0.30	0.0
The power struggle between M&E unit and organizational structure	9.0	8.64	0.67	0.50
Poor utilization of M&E reports and findings	9.0	8.82	0.6	0.0
*Poor coordination	9.0	8.55	1.04	1.0

* *Indicates new factors introduced from Delphi Round One*

poor approach to M&E data collection and analysis, weak linkage between planning and M&E (Seasons, 2003), power struggle between M&E unit and organizational structure (Muriithi & Crawford, 2003), lack of established M&E units (Cameron, 1993) and poor communication (Diallo & Thuillier, 2005) revealed similar impacts of these challenging attributes on the success of M&E in the Ghanaian construction industry. Hence, all 12 challenging attributes listed to assess the impact of M&E in the GCI were rated to have a high impact based on their median scores of between 7 and 8. The challenging attributes also achieved consensus, obtaining IQD ≤ 1.

12.4.4 Impact of effective M&E determinants on the success of project delivery in the GCI

The impact of M&E determinants, namely stakeholder involvement, budgetary allocation and logistics, political influence, technical capacity and training, approach to M&E, effective leadership and effective communication on project success were assessed. Eleven project success (PS) factors were assessed for the impact M&E has on them. Consensus was largely evident on all 11 project success indicators when the assessment of M&E determinants of project success was undertaken by experts. A high impact rating was observed (HI: 70–8.99) along with a 100% consensus reached on all project success indicators (Table 12.11). The Delphi study suggests that value for money, client satisfaction, conformity and completion of project on time noted by other studies from different cultural contexts (Papke-Shields, Beise, & Quan, 2010; Chin, 2012; Ika Diallo, & Thuillier, 2012; Chipato, 2016) shared similar characteristics with the Ghanaian construction industry. This was evident as experts rated all project success indicators as having a high impact based on the observed median score.

Table 12.11 Project success indicators

Indicators	Median	Mean	Std. Dev.	IQD
Completion of project on time	8.0	8.00	0.58	0.0
Achieving conformity	8.0	7.57	0.79	0.50
Meeting project cost	8.0	8.14	0.69	0.50
Stakeholder satisfaction	8.0	8.00	0.58	0.0
Contractor performance	8.0	7.86	0.69	0.50
Health & safety performance	7.0	7.43	0.53	1.0
Value for money	8.0	7.86	0.69	0.50
Environmental performance	8.0	7.43	0.79	1.0
End user satisfaction	8.0	7.86	0.69	0.50
Client satisfaction	8.0	8.00	0.58	0.0
Fit for purpose	8.0	7.43	1.13	0.50

Summary

This chapter presented the results and discussion of the two-round Delphi study. Also presented are the demographic characteristics of the experts. Ninety-eight attributes categorized into seven main factors achieved consensus at the end of the second round of the Delphi study. These main and sub-attributes would influence the modification of the holistic conceptual IM&E model for the Ghanaian construction industry. It was, however, indicative that the Ghanaian construction industry lacks a policy framework for project delivery and, as such, there is an urgent need for an M&E policy framework for the Ghanaian construction industry. It is also confirmed from the study that M&E practice in the Ghanaian construction industry is influenced by the involvement of key stakeholders, the M&E budget allocation, technical capacity, communication and the leadership role. Consensus was achieved on some challenges outlined amongst the 11 experts empanelled. Regarding the relationship between M&E and project success, experts largely agreed that there is a significant relationship suggesting effective M&E will lead to project success.

References

Abrahams, M. A. (2015). A review of the growth of monitoring and evaluation in South Africa: Monitoring and evaluation as a profession, an industry and a governance tool. *African Evaluation Journal*, 3(1), pp. 1–8.

Barasa, R. M. (2014). *Influence of monitoring and evaluation tools on project completion in Kenya: A case of Constituency Development Fund projects in Kakamega County, Kenya.* Kenya: University of Nairobi.

Bikitsha, L., Mamafha, K. & Ngomane, N. (2014). Understanding the use of emotional intelligence during the project leadership process: A case of project managers. *Journal of Leadership and Management Studies*, 1(1), pp. 5–16.

Cameron, J. (1993). The challenges for monitoring and evaluation in the 1990s. *Project Appraisal*, 8(2), pp. 91–96, doi:10.1080/02688867.1993.9726893

Chin, C. M. M. (2012). *Development of a project management methodology for use in a university-industry collaborative research environment.* University of Nottingham.

Chipato, N. (2016). *Organisational learning and monitoring and evaluation in project-based organisations.* Stellenbosch: Stellenbosch University.

Diallo, A. & Thuillier, D. (2005). The success of international development projects, trust and communication: an African perspective. *International Journal of Project Management,* 23(3), pp. 237–252, doi:10.1016/j.ijproman.2004.10.002.

Hardlife, Z. & Zhou, G. (2013). Utilization of monitoring and evaluation systems by development agencies: The case of the UNDP in Zimbabwe. *American International Journal of Contemporary Research,* 3(3), pp. 70–83.

Hwang, B. G. & Lim, E. S. J. (2013). Critical success factors for key project players and objectives: Case study of Singapore. *Journal of Construction Engineering and Management,* 139(2), pp. 204–215.

Ika, L. A., Diallo, A. & Thuillier, D. (2012). Critical success factors for World Bank projects: An empirical investigation. *International Journal of Project Management,* 30(1), pp. 105–116, doi:10.1016/j.ijproman.2011.03.005.

Iqbal, N., Anwar, S. & Haider, N. (2015). Effect of leadership style on employee performance. *Arabian Journal of Business and Management Review,* 5: 146. pp. 1–6 doi:10.4172/2223-5833.1000146

Kamau, C. G. & Mohamed, H. B. (2015). Efficacy of monitoring and evaluation function in achieving project success in Kenya: A conceptual framework. *Science Journal of Business and Management,* 3(3), p. 82, doi:10.11648/j.sjbm.20150303.14

Kimweli, J. M. (2013). The role of monitoring and evaluation practices to the success of donor funded food security intervention projects: A case study of Kibwezi District. *International Journal of Academic Research in Business and Social Sciences,* 3(6), p. 9.

Kolzow, D. R. (2014). *Leading from within: Building organizational leadership capacity.* International Economic Development Council, pp. 1–314.

Luthra, A. & Dahiya, R. (2015). Effective leadership is all about communicating effectively: Connecting leadership and communication. *International Journal of Management & Business Studies,* 5(3), pp. 43–48.

Mugambi, F. & Kanda, E. (2013). Determinants of effective monitoring and evaluation of strategy implementation of community-based projects. *International Journal of Innovative Research and Development,* 2(11). pp 67–73.

Mugo, P. M. & Oleche, M. O. (2015). Monitoring and evaluation of development projects and economic growth in Kenya. *International Journal of Novel Research in Humanity and Social Sciences,* 2(6), pp. 52–63.

Muiga, M. I. J. (2015). *Factors influencing the use of monitoring and evaluation systems of public projects in Nakuru County.* Kenya: University of Nairobi.

Mulandi, N. M. (2013). *Factors influencing performance of monitoring and evaluation systems of non-governmental organizations in governance: A case of Nairobi, Kenya.* Kenya: University of Nairobi.

Muriithi, N. & Crawford, L. (2003). Approaches to project management in Africa: Implications for international development projects. *International Journal of Project Management,* 21(5), pp. 309–319, doi:10.1016/S0263-7863(02)00048-0

Musomba, K. S., Kerongo, F. M., Mutua, N. M. & Kilika, S. (2013). Factors affecting the effectiveness of monitoring and evaluation of constituency development fund projects in Changamwe Constituency, Kenya. *Journal of International Academic Research for Multidisciplinary,* 1(8), pp. 175–216.

Ogolla, F. & Moronge, M. (2016). Determinants of effective monitoring and evaluation of government funded water projects in Kenya: A case of Nairobi County. *Strategic Journal of Business & Change Management,* 3(1), pp 329–358.

Patton, M. Q. (2003). Inquiry into appreciative evaluation. *New Directions for Evaluation*, 2003(100), pp. 85–98.

Papke-Shields, K. E., Beise, C. & Quan, J. (2010). Do project managers practice what they preach, and does it matter to project success? *International Journal of Project Management*, 28(7), pp. 650–662, doi:10.1016/j.ijproman.2009.11.002.

Popa, B. M. (2012). The relationship between leadership effectiveness and organizational performance. *Journal of Defense Resources Management*, 3, 1(4), pp. 123–126.

Seasons, M. (2003). Monitoring and evaluation in municipal planning: Considering the realities. *Journal of the American Planning Association*, 69(4).

Stitt-Gohdes, W. L. & Crews, T. B. (2004). The Delphi technique: A research strategy for career and technical education. *Journal of Career and Technical Education*, 20(2), pp. 55–67.

Van Mierlo BC. 2011. Approaches and methods for monitoring and evaluation. Syscope Magazine. 2011(summer): 31–33.

Waithera, L. & Wanyoike, D. M. (2015). Influence of project monitoring and evaluation on performance of youth funded agribusiness projects in Bahati Sub-County, Nakuru, Kenya. *International Journal of Economics, Commerce and Management*, 3(11), pp. 375–394.

Windapo, A., Odediran, S. & Akintona, R. (2015). Establishing the relationship between construction project managers' skills and project performance. Proceedings of the 51st Annual Conference of the Associated Schools of Construction, 22-25 April 2015, Texas A& M University, College Station, Texas.

Index